MAR 0 5 2013

Net Smart

Net Smart

How to Thrive Online

Howard Rheingold

drawings by Anthony Weeks

302.23
RHE
09/12

The MIT Press
Cambridge, Massachusetts
London, England

MIT Press books may be purchased at special quantity discounts for business or sales promotional use. For information, please email special_sales@mitpress.mit.edu or write to Special Sales Department, The MIT Press, 55 Hayward Street, Cambridge, MA 02142.

This book was set in Stone Serif and Stone Sans by the MIT Press. Printed and bound in the United States of America.

Library of Congress Cataloging-in-Publication Data

Rheingold, Howard.
Net smart : how to thrive online / Howard Rheingold ; drawings by Anthony Weeks.
    p.   cm.
Includes bibliographical references and index.
ISBN 978-0-262-01745-9 (hbk. : alk. paper)
1. Internet—Social aspects. 2. Information technology—Social aspects. 3. Electronic information resources. 4. Social media. 5. Digital media. I. Title.
HM851.R52   2012
302.23'1—dc23
2011040573

10 9 8 7 6 5 4 3 2 1

For Judy

Contents

5   Social Has a Shape: Why Networks Matter   191

6   How (Using) the Web (Mindfully) Can Make You Smarter   239

Acknowledgments

This book has been, even more than usual, dependent on networks and communities of colleagues as well as helpers. First thanks go to Fred Turner, who nudged me to write a book about social media literacies. The idea would have remained just that without belief, support, guidance, and diligent work by my agent, Lydia Wills. Thank you so much for your patience and steadfast belief in this project, Lydia.

Bob Prior has believed in me for longer than any editor in my career and is responsible for keeping two other books of mine in print, in addition to this one. Thank you, Bob. I am grateful to copy editors Deborah Cantor-Adams and Cindy Milstein, who have been a pleasure to work with.

My research assistants were numerous and prodigious: Brian Edgar, Catherine Faas, James Jacobs, Johan Jessen, David Kudler, Maxine Litre, Joey Mornin, Christopher Neal, So-Eun Park, Stephanie Parker, Kit Richert, Jason Schultz, Michelle Wolverton, and Mandy Zibart. My appreciation to all for such excellent work.

The Institute for the Future both supported and partnered with me in my investigation of cooperation studies. I am thankful for the support of Paul Saffo, Andrea Saveri, Kathi Vian, and Marina Gorbis.

Mimi Ito and Cathy Davidson deserve special mention for their ongoing support and mentoring.

My illustrator, Anthony Weeks brought this work out of the text dimension and into the mind's eye. Thank you, Anthony.

My wife was my first reader, as she has been for my entire career. "Thank you," is inadequate, Judy. My daughter, Mamie, was sometimes my native guide, sometimes my ideal reader, and sometimes the critical eye I needed. I appreciate your feedback more than you know, Mamie.

Introduction: Why You Need Digital Know-How—Why We All Need It

The future of digital culture—yours, mine, and ours—depends on how well we learn to use the media that have infiltrated, amplified, distracted, enriched, and complicated our lives. How you employ a search engine, stream video from your phonecam, or update your Facebook status matters to you and everyone, because the ways people use new media in the first years of an emerging communication regime can influence the way those media end up being used and misused for decades to come.[1] Instead of confining my exploration to whether or not Google is making us stupid, Facebook is commoditizing our privacy, or Twitter is chopping our attention into microslices (all good questions), I've been asking myself and others how to use social media intelligently, humanely, and above all mindfully. This book is about what I've learned.

I believe that learning to live mindfully in cyberculture is as important to us as a civilization as it is vital to you and me as individuals. The multifold extension of human minds by chips and nets in the first decade of the twenty-first century has granted power to billions, but in these still-early years of multimedia production studios in your pocket and global information networks in the air, it is clear to even technology enthusiasts like me that our enhanced abilities to create and consume digital media will certainly mislead those who haven't learned how to exert mental control over our use of always-on communication channels.

The mindful use of digital media doesn't happen automatically. Thinking about what you are doing and why you are doing it instead of going through the motions is fundamental to the definition of mindful, whether you are deciding to follow someone on Twitter, shutting the lid of your laptop in class, looking up from your BlackBerry in a meeting, or consciously deciding which links *not* to click. Although educational institutions have been slow to incorporate digital literacies, practical know-how is available to those who figure out how to find it. This know-how, from the art of growing social capital in virtual communities to the craft of cultivating wiki collaboration, might determine whether life online will drive us to distraction, or augment and broaden our minds.

For individuals, the issue of where digital culture may be heading is personal as well as philosophical: knowing how to make use of online tools without being overloaded with too much information is, like it or not, an essential ingredient to personal success in the twenty-first century. Just as learning to drive an automobile (or at least learning how to survive as a pedestrian) was crucial for citizens of the early twentieth century, learning how to deploy attention in relation to available media is key today for success in education, business, and social life. Similarly, those who understand the fundamentals of digital participation, online collaboration, informational credibility testing, and network awareness will be able to exert more control over their own fates than those who lack this lore.

I see a bigger social issue at work with digital literacy, in addition to personal empowerment: if we combine our individual efforts wisely, enough of the right know-how could add up to a more thoughtful society as well as enhance those individuals who master digital network skills. Web 2.0 impresario Tim O'Reilly claims that the secret sauce behind Google, Wikipedia, and the Web itself is the "architecture of participation," enabling countless small acts of self-interest like publishing a Web page or sharing a link to add up to a public good that enriches everybody. Examples of the social-media-enabled public goods that grow out of self-interested actions include the Web and free online search engines.[2]

I don't believe that technology itself, a fixed human nature, or the powers that be wholly determine who ends up in control and who ends up being controlled by others when a communication medium is adopted. But I do recognize that powers eventually emerge that try to close gates, meter resources, and lock down liberties. I'm enough of an optimist to persist in believing that this hasn't happened quite yet, despite real advances in the direction of control by governments and corporations around the world. Right now (and for a limited time), we who use the Web have an opportunity to wield the architecture of participation to defend our freedom to create and consume digital media according to our own agendas. Or by not acting in our own interests, we can let others shape our future.

If I am correct that informed actions might still influence the outcome, declaring that technology alone will solve social problems caused by the use of technology is dangerously naive; at the same time, it is dangerously nihilistic to dismiss all the mental and social tools that microchips make possible as irredeemably destructive. People's actions influenced the ways print media shaped the cultural evolution of the past five hundred years.[3] The early users of the telephone insisted on using it to socialize, not as the broadcast medium envisioned by the first telephone companies.[4] Just as people in previous eras appropriated printing presses and telephones in ways that the inventors and vendors of the enabling technologies never imagined, the shape of the social, economic, political, and mental infosphere now emerging from the combination of inexpensive though powerful computers, mobile communication devices, and global digital networks is not yet fully hardened, and thus can still be influenced by the actions of literate populations. We're in a period where the cutting edge of change has moved from the technology to the literacies made possible by the technology.

Five hundred years ago, Gutenberg presses did not immediately enable people to overthrow monarchies, drive the Protestant Reformation, and invent science as a collective enterprise. The interval between the technological advance of print and the social revolutions it triggered was required for literacy to spread. Print, a technology that leverages the power of the human mind by making possible mass distribution of written documents, required decades for the intellectual skill of decoding those printed pages to spread through populations. The sheer scarcity of painstakingly crafted manuscripts (the word manuscript literally means "written by hand") had constrained literacy for thousands of years. Thirty thousand pen-and-ink books existed in all of Europe in Johannes Gutenberg's lifetime, but more than ten million printed books became available within fifty years of his invention.[5] The sudden abundance of printed material meant that the mental know-how that had been reserved for elites for millennia abruptly became available to anybody who was able to put in the effort to learn to read. For decades and centuries after Gutenberg, newly literate populations began to learn what to do with the new media of their time, and then they started to foment the Reformation, institute political self-governance, and systematize the discovery of knowledge.

Digital literacies can leverage the Web's architecture of participation, just as the spread of reading skills amplified collective intelligence five centuries ago. Today's digital literacies can make the difference between being empowered or manipulated, serene or frenetic. Most important, as people who are trying to get along day to day in a hyperscale, warp-speed civilization that seems so often to be beyond anyone's control, digital literacy is something powerful we can learn as well as exercise for ourselves and each other.

Who Needs to Read This Book, and Why?

I know from my own thirty years online and quest to learn from people who are highly skilled in the new media that practical know-how does exist and can be useful (maybe even essential) to:

• Adults who are adept at using online tools and networks, but face challenges of time and attention management, and seek a balance between their physical and virtual environments

• Intelligent but perhaps less knowledgeable and fearful parents of young people who are going online for the first time, or spending more and more time online

• Young people who are immersed in the digital "hanging out, messing around, and geeking out" online that is such an important part of youth culture today, but are ready to learn deeper, broader ways of using social media productively and collaboratively[6]

• People who are old enough to remember the world before it was webbed, and are simultaneously puzzled, attracted, and fearful about new media

• Businesspeople who want their employees to be net smart with each other inside their enterprise as well as social media literate when dealing with customers—net smarts within enterprises are different from social marketing competencies

• Educators who want to help students connect old and new literacies, and think critically about their own media use

While we're waiting for research to provide more definitive evidence about what our media practices are really doing to our minds and social relationships, I think we can all benefit from adopting some of the rules of thumb discovered by mindful digital media users. Literacy as I am using the term is definitely a skill. But solitary skills are not enough today. Literacy now means skill plus social competency in using that skill collaboratively. Learning how to ride a bike is a skill you have to learn alone, and even if you're the only person in the world who can ride a bicycle, you could get from place to place faster because of your operational knowledge, along with a working bicycle. If you are the only person in the world who knows how to read, write, or hyperlink, however, your skill is far less useful than it could be. What matters the most with present-day new literacies are not just the encoding and decoding skills an individual needs to know to join the community of literates but also the ability to use those skills socially, in concert with others, in an effective way.

I want to introduce you to new know-how (and how to know in new ways) by sharing what I've learned about five literacies that are in the process of changing our world: attention, participation, collaboration, the critical consumption of information (aka "crap detection"), and network smarts. When enough people become proficient at these skills, then healthy new economies, politics, societies, and cultures can emerge. If these literacies do not spread through the population, we could end up drowning ourselves in torrents of misinformation, disinformation, advertising, spam, porn, noise, and trivia. Information overload only begins to describe the problem facing everyone with an email account. The free flows of information that digital technologies have made possible are enriching if used properly, but unhealthy for us as individuals, unproductive for businesses, and toxic for our societies if we don't know how to take them in (or selectively shut them out), evaluate and assimilate them, and contribute our own participation or collaboration—and perhaps most important, when and why to turn off the device, or tear ourselves away from it.

We need to handle the new flows of knowledge, media, and attention in a healthy, flexible, grounded manner, whether we are older and trying to cope with a world that has changed on us, or just starting out in an era in which the rules are still being written. The well-being of sixteen year olds, sixty year olds, start-up companies, and global corporations increasingly depends on the same know-how and how to know.

How Our Learning Journey Will Proceed

In the chapters that follow, I share specific advice about benefiting from and protecting yourself from today's always-and-everywhere media. I direct this advice to worried parents, anxious and enthused students, concerned teachers, curious managers, ambitious employees, thoughtful entrepreneurs, reflective online enthusiasts, puzzled policymakers, and technoskeptics who are just trying to cope. If you need to know what to tell your children about life online, need help surviving and thriving in your own online life, or are grappling with the changes that always-on media are bringing to your organization, I offer the following stories, advice, arguments, evidence, tools, and exercises for your use. I offer this book to people of any age who are willing to think for themselves about their part in digital culture.

I can't give you what you need, however, without some work on your part, precisely because you know better than I do about who you are and where you stand. I can only point out what I've learned and what others

have discovered, and leave it to you to make decisions according to your own values. Here, I strive for a balanced approach that is neither a techno-utopian sales pitch nor a neo-Luddite moral panic; it is instead a pragmatic stance that takes into account the reality that the preferences and circumstances of each reader will differ.

As one of the earliest adopters of what I called "mind amplifiers" (in 1985)[7] and the person who gave a name to "virtual communities" (in 1987),[8] I have learned that the media I've been using with gusto for three decades also have their downsides. Although I've traveled across countries and disciplines to consult with a wide variety of media experts, much of what I convey here in terms of practical advice comes from my own experience. I've learned to be wary of trying to sell others the generalizations about life online that I've found to be true through my own exploration—because one of the things I've learned about social media is that the same activity can be a lifeline for one person and a distracting compulsion to others. There is no single recipe for a mindful life in the digital mediasphere; reflection is required.

One tool that I do feel comfortable generalizing about is the importance of questioning my own communication practices—recognizing which media and mediated social activities I tend to avoid, which ones attract or distract me, and which lead and mislead me, and reflecting on *why* I react in these ways. I have found through years of trial and much error that the most enriching, least harmful way for me to live in my own computer-mediated world is to cultivate an occasional but ongoing inner inquiry into whether my own activity of the moment is really as significant as what is happening in the rest of my life at each moment. You can't make microdecisions about how to deploy your attention in the moment unless you have made macrodecisions about how you want to spend your time. And while I'm asking questions, where is my body while my mind scurries through cyberspaces? It's easy to ask oneself, What do I think I should be doing right now? Answering it usually takes work. The process of trying to address the question in your own context is the work of learning digital mindfulness.

Each of the five literacies I discuss is connected to and in many cases undergirds each other. It's impossible to separate signal from noise without exercising attention, so mindfulness is a prerequisite to effective crap detection. Similarly, it's difficult to instigate mass collaboration without network awareness, nor is it easy to participate online without also collaborating. Twitter is a recent example of a social medium that can be a waste of time or multiplier of effort for the person who uses it, depending on how knowledgeable the person is in the three related literacies of attentional discipline, collaborative know-how, and net savvy. You need to know who to pay attention to when you "follow" other Twitter users, how to participate in the networks of trust and norms of reciprocity among Twitter users that make for social capital, and how to craft messages that others will propagate to their own networks. Attention is a literacy that can thread all the other literacies together and hence is fundamental to the others in several ways, so I'll start there.

In the first chapter, I connect my own experience, the exercises recommended by others, and what I've learned about the underlying neuroscience of attention to the practical literacy of controlling attention. The learning journey here begins with an updated understanding of how attention works, why distraction and multitasking might or might not be the vehicle through which modern media are making us stupid as individuals and shallow as a culture, and then gets right into what to do about the dangers of distraction through examining mindfulness, ancient and modern. I'll lay the foundations for discussions later in the book about the possibilities of the extended mind—the use of technology to go beyond remedies for attentional deficits to methods of enhancing intellectual performance. Most crucial for you and your power to wield the literacies introduced later, the first chapter will demonstrate how to begin to take control of your most important technological affordance—your attention.

In the second chapter, I'll show how to use your attention and mine, individually and in concert, to filter out the noise and crap in order to concentrate on the tiny relevant portion of the moment-to-moment incoming tsunami of information. Intention added to attention, and mixed with knowledge of information-filtering tools, work together in a coordinated mind-machine process I call "infotention." Critical thinking, information filtering, and Ernest Hemingway's fundamental "internal crap detector" are all about how to use your attention to begin managing the inflow of media. Like the first chapter, my exploration of search and credibility skills as well as attitudes is about the meeting of mental capabilities with the technologies of keyboards, screens, and networks. Together, the first two internally focused chapters are about what my friend Cathy Davidson, educational technology pioneer, calls "your brain on the Internet."[9]

Moving from the strictly individual mental aspects of life online to the coupling of individual personality with digital culture via social media, the third chapter is about the literacy of participation, or the know-how that empowers the best of bloggers and videobloggers, netizens, Twitterers, and online community participants—those who use digital media to express themselves, socialize, advocate, organize, educate, and grow collective intelligence. Mirroring the inner-outer powers covered in other chapters, participation is about internal individual skills and strategies, and at the same time, the Webwide aggregation of participation—where the literacy of participation shades into the literacy of collaboration. A "participatory culture," as media analysts Henry Jenkins and Mizuko Ito put it, is one in which the level of digital participation—from gaming to curating—creates a social setting in which citizens become active agents in cultural production.[10] Conversely, if the level of participation literacy fails to maintain a certain (presently unknown) minimum, a social setting for media use in the future might hark back to the mediasphere of the broadcast age, in which a relatively small population of prosperous, empowered producers broadcast their versions of culture to a much larger, far more passive and less wealthy, less powerful population.

Chapter 4 moves from the personal and interpersonal to the cybersocial. The know-how at the core of this literacy is about the magic of several different flavors of collaboration made possible by networked media. The realms of collaboration are broad and deep, so this chapter offers both a high-altitude map of the territory of online collaboration and close-up conversations with the people who have created famously successful collaborative enterprises. Wikipedians, Flickr taggers, and social bookmarkers are contributing new knowledge in new ways by performing self-interested information practices within an architecture of participation that provides value to all. Virtual community organizers work at the border of media and interpersonal relationships, in a zone where technical knowledge will get you nowhere if you don't understand online social norms—and can get you much that money can't buy if you know how these emerging cybersocial forms work. As one of the earliest commentators on cybersociality, I can speak from experience about the benefits and pitfalls of mediated communities.

Collective activities and interpersonal capabilities that nobody dreamed possible have become part of everyday life for millions of people. In 1985, when I participated in an ad hoc online support group for a member of our virtual community whose son had a life-threatening disease, we did speculate that this kind of group might be used in the future by more than the early adopters of networked social communication. Until my "Virtual Communities" article in 1987, there really wasn't a word or cultural category for strangers who lived in different places yet offered each other sympathy, medical advice, and even financial support.[11] Millions use services today such as Patientslikeme.com and mdjunction.com.[12]

Knowledge creation, political activism, and health support are far from the only ways people are working together with others they have not been able to work with before, in ways and places that were never before possible. Online collaboration may be evolving a third variety of economic production to supplement the market and firm, as scholars such as Harvard's Yochai Benkler contend: "Who could have predicted that volunteers, working with neither financial incentives as we know them or the management structure of the firm as we know it managed to co-create free, open source software that challenged Microsoft in both the operating system and web browser markets?"[13] A coalition of volunteers who build and improve millions of articles in hundreds of languages as part of a free encyclopedia would have sounded preposterous even to enthusiasts when the Web first became widely known in the mid-1990s. Today, succeeding online—in business, personal life, and the public sphere—can entail knowing how to find, participate in, and grow your own virtual community.

"Collective intelligence" and "crowdsourcing" are other emergent terms to capture newfangled forms of collaboration. People who don't communicate directly as they do in virtual communities can nevertheless aggregate individual efforts to create useful public goods. By bookmarking and tagging Web sites that contain useful information, people are creating a kind of mass-curated knowledge that would have been impossible before the Web. The Library of Congress, lacking the funds to exhaustively describe its photographs of U.S. life, put them up on Flickr, where volunteers tagged millions of them—for no financial return.[14] Future forecasters are beginning to use the voluntary, enthusiastic, communal efforts of online gamers to foresee and attempt to solve world-scale problems. New ways for people to collaborate are invented on an ad hoc basis every day. For example, when computer scientist Jim Gray went missing at sea, his friends obtained recent satellite images of that ocean region from NASA and Google; Microsoft and Amazon engineers divided the images into a half-million separate pictures; more than twelve thousand volunteers searched the photos. Gray was never found, but a new kind of crowdsourcing popped into public consciousness.[15]

Chapter 5 is about the multifaceted knowledge of networks that comes in handy so often today. Network savvy is exceptionally multidisciplinary. Becoming network aware has to include some basic knowledge gained by sociologists who have studied the way structural dynamics of networks influence how people relate via social networks; another bit of sociology, the famous "small-world network" that explains how every human being is connected to every other human, applies directly to online network building. Now that more than half a billion people have their own Facebook pages and more than five billion carry mobile phones, sociologists have also been tracking a shift of central importance to digital citizens: the emerging phenomenon of "networked individualism." Political scientists and sociologists alike have been the specialists who use the term "social capital" to describe the power of populations to get things done together outside formal laws and institutions, but now every knowledgeable Web user needs to understand how online behavior can grow or obstruct social capital. Small worlds, networked individuals, and social capitalists are all part of the emerging culture of digital publics. The politics and psychology of privacy—and why knowledge of privacy protection is critical in an era of transparency—is another case of an issue that small groups of specialists debated a decade ago, but now poses daily challenges for parents, students, and citizens. It's hard to be much of anywhere in the twenty-first century and not recognize what University of Southern California professor Manuel Castells succinctly argues: that networks matter.[16]

The shapes of our connections and what we know about them are not only the subject matter of a developing new science of networks but also matter in the ways that technological networks amplify and extend human social networks. Technological architectures and the media practices of ordinary people suddenly matter very much in the personal realms of liberty, opportunity, and the possibility of justice.

By the end of chapter 5, you should have a set of mental and social tools to apply to your own advantage—and the benefit of others. You'll have the knowledge to confront the bigger question of what social media mean cognitively and socially. The final chapter frames these practical literacies in relation to the broader issues of privacy, remix culture and copyright conflicts, and the role of today's citizens in the digital public sphere. It also provides advice for parents and a bullet-point summary of our learning journey.

Attention! The Fundamental Literacy

Last month, I picked up my twenty-six-year-old daughter at work to take her dinner. Here is our conversation, verbatim:

Me: "Honey, is it necessary for you to spend our entire time together on your BlackBerry?"

She: "Daddy, if I don't deal with today's work emails before dinner, I'm going to fall behind."

Me: "Welcome to my world. I think."

That exchange started me pondering previously somewhat-separate issues that had been on my mind. On the day I met up with my daughter, my university students and I had been contending with the attention issues raised by their use of laptops in class. I can remember when "you've got mail" was, for most people, a cheery and inviting message. Now I was seeing how my daughter was already on a digital treadmill I know well, and I know that when I travel, I fear the hundreds of messages in my in-box when I return. At home, my wife frequently has to finish her online messaging before she turns away from her computer to greet me when I walk in the door. I know that I have been guilty of the same kind of social media delay in face-to-face sociability with the most important person in my life; I often have to finish my email, instant message, tweet, text, bookmark, tag, post, or comment before I greet my wife when she walks in the door. And long before our daily lives were colonized by pocket-size communication gadgets, I regularly wondered why my whole family jumped to answer a ringing telephone when we knew the caller would roll over into voice mail if we chose not to interrupt our dinner conversation.

These concerns are not unconnected, of course. And neither are they wholly new in a larger sense, although novel social media behavioral challenges seem to pop up every day. We've been reallocating our attention in response to new communication media for a long time.

Once I started looking for everyday behaviors where communication technology use affects attention, it didn't take long to perceive the outlines of the large-scale shifts in attentional practices and norms that we all see happening around us in many ways. I understand my daughter's fear of the overflowing email in-box, as do most white-collar adults in the industrialized world. I talk to my wife while she is texting (and vice versa). You don't have to wear a white collar to have sent one of the trillions of text messages transmitted worldwide this year. My daughter had six instant messaging windows open while she chatted on the phone and worked on a school assignment (and I let her do it because she is an excellent student, so maybe I helped set the stage for her BlackBerry habit). I see how all eyes in my university classroom are not on me but rather on laptop screens. I started noticing myself at the same time I was observing the attentional behaviors of others; I began to think in new ways about how people deploy their attention when I started looking at the way my own thinking processes had changed since I turned my typewriter in for a personal computer (PC), and then plugged my PC into a modem and thus my first online network.

Although I originally started using digital tools in order to type more efficiently, I soon learned that the transition from electric typewriter to word processor entailed more than just a change in office machinery. Not long after I began using computers and networks, I started writing about how it felt to use them. The ways in which our uses of social media affect our minds, relationships, and society have been the overarching theme of the books I've written for the past twenty-five years. I started to teach courses on social media five years ago because I recognized the importance of helping students examine their own psychological and social issues around digital media use. Teaching and learning with students in classrooms at Berkeley and Stanford brought me into direct contact with (and provided a living laboratory in) generational differences in attention patterns.

Probably the first advice I would give unequivocally, based on my own decades online, is that in a world where information is abundant and veracity is not guaranteed, while gatekeepers, authorities, and fact-checkers are scarce, each of us as individuals and all of us as a society have no choice but to learn how to think critically about what we pluck from the information flow, how much we are to believe what we find or are given, and whether we should even devote any mind share to it at all.

Although I hope to explain what is known about the cognitive effects of using digital networked media, including the research and controversy over multitasking, this is not a book on multitasking, pro or con, or how to manage your time better; there are plenty of those. Neither is this book going to deal with the issue of attention deficit disorder. Knowing when as well as when not to multitask is a key part of the digital literacy toolbox—and you don't need to have a disorder to be confused about how to react to rapid social and technological changes. If you aren't a little confused, maybe you aren't thinking deeply enough about the bigger picture. For the purpose of my inquiry into a broad range of literacies, concentrating too much on the important but not all-encompassing issue of multitasking risks missing larger issues about a broad range of attentional habits that are dying and aborning.

Most people in the world recognize, at some level, that a massive shift is taking place in the way we direct, fail to direct, fragment, or time-share our attention in conversations, classrooms, and while walking down the street. Many are uneasy about this transformation. Some, like Nicholas Carr in his article "Is Google Making Us Stupid?" and his book *The Shallows*, believe we are losing an essential ability to focus and dive deep.[17] The sociotechnological questions Carr addresses may have been made possible by the digital devices a majority of the earth's population now carry, but the real changes driving this shift are occurring in human minds and between human beings, not in microchips. The way we communicate today is altering the way people pay attention—which means we need to explore and understand how to train attention now, so that we, not our devices, control the shape of this alteration in the future.

It's not that multitasking is always bad (except when it is—like when you are driving a car), or continuous partial attention (such as surfing the Web while talking on the phone) is always rude and inefficient. It's that too few have learned and taught to others the skills we need to know if we are to master the use of our attention amid a myriad of choices designed to attract us. A significant part of the population has not yet learned to decide when it is appropriate to share multiple lines of attention and when single focal point is necessary (and I'm not just talking about etiquette here but rather about efficacy in business and personal lives), nor have many people studied how attention can be trained. Who can blame us? We've been busy trying to catch up with the way our uses of digital computers, worldwide webs, and mobile cameraphones have restructured our lives. (A 2010 survey found that one in six adults has physically bumped into someone or something while talking or texting on their mobile phone.)[18]

Fortunately, learning to gain control over attention is a skill that people have been perfecting for thousands of years, and it can start with something as simple as paying attention to your breathing. Eventually, twenty-first-century elaborations on older mind tools have to be learned, but the beginner in traditional meditation discipline and modern digital infotention training both start in the same place: elementary mindfulness exercise involving attention to the physical breath.

One of the most critical things to know about mindfulness training is that even the smallest amount of attention is immeasurably more useful than none at all. Step one in gaining control of attention is to simply notice it. Getting started in this kind of reflective thinking is the hardest part, and yet it's also easy to begin. After embarking on what should become at least occasional self-examination, it's time to turn the tool of attention control—however early you might be in your self-training—to the task of finding the information you need at the moment you need it, learning what you need to learn and forgetting what you don't need, and most important, learning how to filter out the bad info.

Calibrating Your Crap Detector: What You Pay Attention to After You Pay Attention to Attention

The answer to almost any question is available within seconds almost anywhere on earth, courtesy of the invention that has altered forever ancient rules about how we discover, store, and classify knowledge: the search engine. People don't just use online search for homework or business intelligence. Search has penetrated to the quotidian details of daily life like finding a plumber or ordering a pizza. With location-aware devices, information is now available that takes into account where you are, what time it is, which direction you are pointing your device, and what your social network thinks about it. If you have a smart phone, you not only can find the nearest place to eat vegetarian cuisine but also find out what other people have to say about the food and service, get visual and vocal directions to your destination from where you are now standing, and view a photograph of what the block you seek looks like. When today's infants grow up, they will be amazed that their parents' generation could ever get lost, not be in touch with everyone they know at all times, and get answers out of the air for any question.

Materializing answers from the air just in time and just in place turns out to be the easy part—the part a machine (a really, really big machine like the Web) can do. The real difficulty kicks in when you click down into your search results. At that point, it's up to you, the human who is using the machine, to sort the accurate bits and the ones that have immediate relevance for you and your circumstances from the ignorantly or maliciously inaccurate information. While our public schools do a poor to fair job preparing students for life in the nineteenth and twentieth centuries, instruction in online search and credibility testing for our current milieu is not taught in most classrooms.

Unless a great many people learn the basics of online crap detection, and begin applying their critical faculties en masse and soon, I fear for the Internet's future as a useful source of credible news, medical advice, financial information, educational resources, and scholarly as well as scientific research. Some critics argue that a tsunami of hogwash has already rendered the Web useless. I disagree. We are indeed inundated by online noise pollution, but the problem is soluble. The good stuff is out there, if you know how to find and verify it. Basic information literacy, widely distributed, is the best protection for the knowledge commons; a sufficient portion of critical consumers among the online population can become a strong defense against the noise-death of the Internet.

The first thing we all need to know about information online is how to detect crap, by which I mean information tainted by ignorance, inept communication, or deliberate deception. Learning to be a critical consumer of Web info is not rocket science. It's not even algebra. Becoming acquainted with the fundamentals of Web credibility testing is easier than learning the multiplication tables. The hard part, as always, is the exercise of flabby think-for-yourself muscles.

Learning how to make use of huge, unsorted, continually changing flows of information without becoming overwhelmed is partly an application of minimally trained attention skills to a simple question in relation to every assertion, factual claim, or opinion: How do I know I should trust this information as accurate? The specifics of examining the credibility of information effectively are as simple as looking for an author's name somewhere on the page in question and submitting it to a search engine, and as complicated as learning to use one's attention in conjunction with the variety of increasingly powerful automated filters that are becoming available. The specific combination of learned attentional skills and learned information technology know-how is an important new aspect of the digital literacy I call infotention.

In the second chapter I introduce an increasingly significant learning tool known as critical thinking that I certainly didn't invent but that has grown to be vitally essential in the many-to-many, anyone-can-publish era. I'll look at how people actually do assess the credibility of what they find online. I'll talk to experts in search and credibility, consider the utility of crowdsourcing your filters, and zoom way back to illustrate how the nature of knowledge, information gathering, and meaning making are changing.

When you're on your way to gaining control of your online attention and have begun to practice crap-detection skills, I turn from "this is your brain on the Internet" to "this is what the Internet enables people to do together." From the cognitive to the social, I'll shift our attention to the technology and sociality of participation and collaboration, focusing on the skills digital citizens need to master in order to take part in or instigate mass collaboration.

By sampling strategies and understanding the benefits of many different kinds of online collaboration, I hope to help you try on collaborative mindsets and ways of using the media freely available to you. Wiki thinking is one form of distributed cognition that has only become possible in recent years. Scholars who indulge in social bookmarking contribute hints about the skill sets ordinary digital citizens ought to have when seeking and trying to make sense of information. Tens of millions of online game players are having fun, and in the opinion of some well-respected business leaders, some of them are honing the collaborative talents essential in knowledge-based enterprises. Huge corporations are crowdsourcing design by asking their customers to help create the products they want. From each of these different milieus, I draw practical lessons regarding the social competencies we need to benefit from the Web's architecture of participation. The following two chapters, on participation and collaboration, introduce the individual and group aspects of collaboration literacy.

What It Takes to Participate in Participatory Culture—and What You Get Out of It

If print culture shaped the environment from which the Enlightenment blossomed and set the scene for the Industrial Revolution, participatory media might similarly forge the cognitive and social environments in which twenty-first-century life will take place. Knowing that you have a printing press, broadcasting station, community hall, marketplace, school, and library of all knowledge in your pocket—and knowing how to use it for your own benefit—is what makes the difference between a consumer of electronic gadgets and an empowered citizen.

Participatory media include every online service that enables individuals to create as well as consume content online. Media as distinctly different as YouTube and World of Warcraft share three characteristics:

• Many-to-many media now make it possible for every person connected to the network to broadcast as well as receive text, images, audio, video, software, data, discussions, transactions, computations, tags, or links to and from every other person. The asymmetry between broadcaster and audience that was dictated by the structure of predigital technologies has changed radically.
• Participatory media are social media whose value and power derives from the active participation of many people. This value derives not just from the size of the audience but also from people's power to link to each other, to form a public as well as a market.
• Social networks, when amplified by information and communication networks, allow for broader, faster, and lower-cost coordination of activities.

People who make even the most modest contributions such as correcting a spelling error on Wikipedia or tagging a photo think of themselves differently from those people who only passively consume the cultural material broadcast by others. A participant is active. A consumer isn't practicing, even in a small way, the skill that is at the foundation of building social capital online—for contributions are often signals to others that it would benefit them to pay attention to and share with you.

The eager adoption of Web-based media by millions of young people around the world demonstrates the strength of their desire—unprompted by adults—to learn digital production and communication skills. According to a 2005 survey by the Pew Internet and American Life Project, "The number of teenagers using the internet has grown 24% in the past four years and 87% of those between the ages of 12 and 17 are online."[19] This interest by U.S. (and Brazilian, British, Chinese, Indian, Japanese, Persian, etc.) youths in media production practices might well be a function of adolescents' needs to explore their identities and experiment with social interaction—and can be seen as a healthy active response to the hypermediated environment they've grown up in.

Whatever else might be said of teenage (and any age) bloggers, dorm-room video producers, or the millions who maintain pages on social network services like Myspace, Facebook, and Google+, it cannot be asserted that they are passive media consumers. They seek, adopt, appropriate, and invent ways to participate in cultural production. Another recent Pew study found that more than 50 percent of today's teenagers have created as well as consumed digital media.[20] This introductory chapter, then, is for those avid young digital media makers (and their parents and teachers) as well as older Web surfers who want to know how to dive deeper (or at least less shallowly) into what the Web has to offer. I do this in the knowledge that addressing the needs of those who are not able to participate in cultural production—the other half of the digital divide—is still an important task. Although significant barriers remain in regard to economically marginal youth and adults, the knowledge and advice in this chapter is geared toward the educational needs and opportunities of the hundreds of millions of people around the world, of many nationalities and socioeconomic levels as well as all ages, who have access to digital media and networks.

Senator Trent Lott lost his position as majority leader of the U.S. Senate, George Allen lost his election to the Senate, and CBS news anchor Dan Rather was forced to retire, all because of the way informed participants used email, blogs, and other participatory media to organize.[21] Participation is power, and any of the more than two billion people who have Internet accounts can learn to wield that power. This chapter looks at how and why to be an active, informed participant in digital culture, and sets you up for the next interconnected literacy—the art of online collaboration.

Clueing in to Collaboration: Making Virtual Communities, Collective Intelligence, and Knowledge Networks Work for You (and Us)

If I had to reduce the essence of Homo sapiens to one sentence, I'd propose: "People create new ways to communicate, then use their new media to do complicated things together." *Why* we act in concert is the big question. People do things together for a rich mixture of reasons, and Web-based collaboration tools are particularly important in this regard, because wikis and bulletin board systems (BBSs) enable people to collaborate in ways that challenge basic assumptions underlying modern economic theory and contradict older stereotypes regarding human motivations to cooperate.[22] The current story that most people tell each other about how humans get things done focuses on the well-known flavors of self-interest that make for great drama—competitive struggles for survival, power, wealth, sex, or glory. I see the outlines of a new narrative emerging, however, in which competition is still central, but its place on our mental map shrinks a little to make room for new knowledge about cooperative arrangements and complex interdependencies.

Starting with the Web's invention (which its creator refused to patent and insisted on giving to the public domain), and continuing with efforts such as the South-East Asia Earthquake and Tsunami Blog, some significant online social behavior suggests that in addition to financial compensation and other forms of naked self-interest, people do things together for fun, mutual enrichment, the love of a challenge, out of compassion, and because we sometimes enjoy working with others to make something beneficial to everybody.[23] This chapter explains how the Web's architecture of participation makes new forms of collective action possible, asks some of the superstars of mass collaboration how they work their magic, and lays out what I've learned from twenty five years of participation in as well as observation of the online activities now called "social media."

The power of sociality stems from human not technological attributes, but tools are created in order to leverage human attributes; any tool that can help humans overcome barriers to cooperation works because it augments an essentially human skill such as persuasion, education, or collaboration. Online social networks can be powerful amplifiers of collective action precisely because of the specific ways they extend the power of human sociality. This augmentation is often but not always healthy. Any tool that expands human capabilities also makes it possible for some of our nastiest predilections to operate on new scales as well. To be sure, gossip, conflict, slander, fraud, greed, and bigotry are part of human sociality, whether it takes place at the village well or in a virtual world, and those parts of human behavior can be amplified too. But altruism, fun, community, collective action, and curiosity are also parts of human sociality—and I propose that the Web is an existence proof that these capabilities can be artificially extended. Indeed, I agree with those who contend that our species' social inventiveness is central to being human.

The parts of the human brain that evolved most recently, and are connected to what we consider to be our "higher" faculties of reason and forethought, are also essential to social life. This is no accident; it appears that human brains and human social behavior shaped each other's evolution. The neural information-processing required for recognizing people, remembering their reputations, and learning the rituals that remove boundaries of mistrust and bind groups together, from bands to communities to civilizations, may have been enabled by (and driven the rapid evolution of) the brain structure unique to mammals—the neocortex.[24] Humans in particular appear to have evolved brains that are optimized for social activity. Is it any wonder that we're now designing social technologies?

Our immediate primate ancestors left the relative safety of the forest to compete with megapredators and saber-toothed everything on the open savanna. *Homo erectus* couldn't run fast, fly, or emit a stream of stinky fluid. They didn't have claws, fangs, or armor. But a couple hundred thousand years ago, these creatures started to outsmart the merely instinctive pack animals by improving their ways of doing things in groups. They coordinated defense and food gathering, and those who were better at participating in or organizing this coordination—probably by learning some new code like spoken language—passed along more of their genes. Homo sapiens evolved to favor, along with the good looking and strong, the most able communicators, and those who could coordinate or at least abide by cooperative efforts. It pays to keep in mind the biological and historical roots of the human drive to cooperate—and how we've always invented ways to overcome hurdles to cooperation—when studying the modern arts of mass collaboration.

Collective knowledge gathering was one of the capabilities that most excited me when I first wrote about virtual communities in 1987: "If, in my wanderings through information space, I come across items that don't interest me but which I know one of my group of online friends appreciate, I send the appropriate friend a pointer to the key datum or discussion."[25] Now, entire communities exist for the purposes of knowledge sharing and organization, from social bookmarking services such as Diigo.com and Delicious.com, to question-answering forums such as Quora (which calls itself "a continually improving collection of questions and answers created, edited, and organized by everyone who uses it") and Formspring.[26] When I recall the days I used an acoustic modem at 110 bits per second to download glorified library catalog entries, the notion of free search engines, free collaboration tools, and voluntary knowledge-building collectives still seems as science-fictional magical as the hyperspace drive in movies and television shows. They are now indispensable everyday tools for billions of people. Those who know how it's done, as always, gain an edge.

Meet Jane McGonigal, for one, who creates massive multiplayer "alternate reality games" that take place in the physical world as well as cyberspace, involve thousands of people worldwide, and tackle real global-scale problems through playful collective intelligence. Or Wikipedia cofounder Jimmy Wales, who spends most of his time traveling to the physical hubs of Wikipedia communities, getting to know the people who have used an ultrasimple online tool to create a free encyclopedia with millions of entries. Every programmer also knows about Linus Torvalds, who sparked the effort globally to create free and open-source software. Tim Berners-Lee didn't ask permission of any central authority, nor did he require any technology provider to rewire the Internet, when he passed around the code for hyperlinks and Web servers that led to the explosive growth of the World Wide Web—just as Ken Thompson freely spread the UNIX operating system that made the Internet possible by sending out the code (then in the form of big reels of magnetic tape) with the appended note, "Love, Ken."[27] Douglas Engelbart envisioned, invented, and persuaded others to invent what we know as the PC, multimedia, and hypertext because he felt it was his duty to improve people's power to cooperate.[28]

All these superheroes of cybercollaboration knew a few simple things that the rest of us can benefit from learning about, such as how to:

- Create a variety of ways to contribute and give volunteers attractive roles
- Enable self-election where people choose what it is they want to work on
- Give participants platforms to work on together for mutual interest
- Acknowledge contributors
- Make decision making transparent (if not necessarily democratic)

In the chapters to come, I'll share these and other examples of collaboration lore that I've picked up from these virtuosos.

It's possible to master the art of controlling attention while you sit alone in a room, but it isn't possible to participate, collaborate, or crap detect without taking advantage of both social and technological networks. Understanding how networks work is one of the key survival skills of the twenty-first century. The next chapter pulls together network science, sociology, practical Facebooking, and the art of online self-presentation to provide both a framework for thinking about and tools for acting effectively in a networked world.

What You Need to Know about Network Smarts—from Small Worlds to Privacy Settings, from Weak Ties to Social Capital

New knowledge about the nature of networks is essential for getting around in this century because digital data and human communication networks erase barriers and multiply possibilities for one of our most powerful capabilities; our sociality. The science of networks emerged in the 1990s when large amounts of data about all kinds of phenomena, together with computer tools to make sense of this information, enabled scientists of different stripes to recognize common characteristics of networks that shape societies, ecosystems, languages, or online social media. This chapter flies over the wildly interdisciplinary landscape of network studies, zooming in on those features that can inform the behavior of digital citizens.

Some of the new knowledge comes from sociologists who were looking at social networks before the Internet was created. The idea of "six degrees of separation," for example, was popularized through a widely reported sociology experiment by Stanley Milgram that used paper letters in the U.S. mail to demonstrate how each human being is connected to every other human being by a surprisingly small number of steps.[29] Decades later, physicists and sociologists noted that small-world networks manifest in widely separated disciplines. The networks of relations in ecosystems, the relationships between words in a language, and the human networks that people create together wherever they congregate all exhibit similar structural characteristics.[30] More recently, social network analysts have presented evidence for "contagion" in social networks: we appear to be influenced by the behavior of people we don't know directly, but who know someone we do know directly.[31]

Electronic engineers and computer scientists have made major contributions: Sarnoff's, Metcalfe's, and Reed's laws of networks explain the extraordinarily rapid rise in value of business enterprises such as eBay and Facebook—and point to entrepreneurial opportunities for anyone who can come up with a new platform for human group formation.[32] When you learn how to look at them, you'll see how discoveries emerging from this new interdisciplinary science point to real-world knowledge that can be useful to mindful digital citizens.

Manuel Castells, a scholar who studies social aspects of networked media worldwide, argues that networks matter now because new technical networks dramatically multiply the power of age-old human cultural tools of sociality, politics, and economics.[33] Castells's insight is worthy of close attention because one key to the successful use of online social networks lies in understanding how online capabilities can be used to enhance social behavior. The shapes of our interpersonal connections matter, Castells maintains, because of the ways technological networks enable the human social networks that enmesh each and every one of us to work more rapidly, in more settings, across more boundaries, than ever before. When you read about the irate airline passenger who organized a protest on Facebook and prompted Congress to consider a "passenger's bill of rights," or the Iranian government using the Internet to track down and arrest protesters, you are hearing about how the politics of networks affects the lives of ordinary people.[34] Technological architectures and people's networked media practices suddenly matter very much in the personal realms of liberty, opportunity, and the possibility of justice.

I didn't let my child loose on the streets without teaching her about traffic and looking both ways. Similarly, I don't like to see otherwise well-educated people loose in digital culture without knowing something about what makes a small-world network work or why a portfolio of weak ties is important. Networks particularly affect privacy and reputation—the places where our private lives intersect or collide with public knowledge, whether or not we know what to do about it. In previous eras, it may have been true that "it's not what you know but who you know." Today, *how you know who you know* matters as much as who you know, and one of the most valuable traits a person could have in a twenty-first-century organization is a knack for knowing "who knows who knows what."

Net smarts are not just vital to getting ahead; you need this knowledge to keep from falling behind. This caveat may well be an argument that our use of technology has grown way beyond our control, but it seems to be a fact of life whether or not we particularly like the idea. And whether or not we do anything about it, the webbed world is full of information about us that is provided by other people, including their opinions about us—the fact of life we know as reputation. To an individual, reputation is a powerful influence on how well one gets on in life, and it's not wholly controllable by the person it impacts. Human sociality has always been thus; indeed, some social scientists suspect that gossip may have been involved in the transition from primate social grooming to human language.[35] Now, however, instead of whispering at the watering hole or scrawling your name on a bathroom wall, reputation assassins can leave indelible and searchable smears on the Web. It turns out that digital networks can also amplify some of human beings' less laudable social behaviors. The art of "presentation of self" becomes all-important when you are trying to wrest control of your reputation from others. As Microsoft Research digital anthropologist danah boyd put it in a conversation with me: "Today people need to frame their own stories, creating a positive living presence on the Web" as the most effective way to put a positive spin on what search engines turn up around your name.[36]

To groups, social capital is the name for the social agreements and communication networks that allow people to get things done together informally, without state or strictly economic institutions. Social scientists such as Harvard University professor Robert Putnam claim that social capital—the mesh of traditional agreements that enable cooperation, and the networks that carry reputation information and thus lubricate transactions—is a key factor that influences the way one society thrives and another struggles.[37] Now that new kinds of human networks emerge online around mutual interests as well as the traditional community catalysts of physical proximity or sectarian allegiances, and social activities are mediated through Facebook, Twitter, YouTube, and Flickr, new ways to cultivate social capital become available. Ask the people who raised $250,000 from Twitter users in two weeks in 2009 to sponsor clean-water projects in impoverished villages.[38]

Network knowledge derives from a variety of disciplines that had previously not been connected (digital networks and human social behavior), and the skills based on this knowledge include a wide variety of situations. I'll restrict my focus here to knowledge, wherever it is derived, that can be applied directly to mindful life online today. When you grasp the basics of social network analysis, you'll know that growing a diverse personal learning network (PLN, as the enthusiasts call it) often is more useful than having a large, homogeneous social network. If you know how others seek to use your digital footprint to market or track, you have the power to protect your privacy and reputation. If you work in an organization, knowledge of the power of "structural holes" that connect networks can help you position yourself as a profitable conduit for good ideas.

None of this knowledge is especially difficult to understand or put into practice. It's just that until now, those of us who want to use network smarts to thrive in digital culture without losing our humanity have had to put the puzzle pieces of theory, practice, and lore together for ourselves. This book is an early attempt to bring these connected but disparate pieces of knowledge together, and surely (I hope) won't be the last. In the future, basic network literacy ought to be a part of school curriculum.

When I started thinking about the relationship between my personal networked media practices and my own thinking and attention, I realized that I've been thinking about thinking tools—how they work, what they mean, and how do I get my hands on them—for decades now. Understanding how present-day PCs and networked media originated in the mid-twentieth century establishes a good foundation for twenty-first-century skills; besides, it's a fascinating story. The tools I used to write this book, and possibly the medium by which you heard about it, grew out of the dreams of people who specifically wanted to use computers to "augment human intellect" and link personal mind amplifiers into an "intergalactic network."[39] These dreamers weren't in the mainstream of the computer industry or computer science. They were people who wanted digital thinking tools for their own use as well as the common good, and set out to create them, even though conventional wisdom held that digital computers were for scientific calculation and business data processing—payrolls and the like.[40] My own career as a writer and teacher was powerfully affected by encounters with some of the people who created the first PC and computer-mediated communication network.

Before I ask you to take my word about what I've learned, I think you'll benefit from hearing how I learned what I propose to teach you.

Dreaming of Mind Amplifiers: A Personal Journey

By the time the first PCs came along, I had been spending my days facing a typewriter and blank page for nearly a decade. I knew little about computers, but I was always interested in the future of media. And it didn't seem too far-fetched to think about using electronics within my lifetime for what Peter Drucker later called "knowledge work."[41]

Since the 1970s, I had been intrigued by the idea that computers—most of which were still programmed through punched paper cards—could be accessed through telephones. What if I could go back to the library multiple times a day, I mused, by plugging my telephone into the library's computerized database—a feat that wasn't possible for me then? I had been tracking the "videotext" experiments that big publishers along with broadcasters like Knight-Ridder and Warner were experimenting with: a soon-to-come way of delivering customized information to people in their homes by using telephones as input devices and televisions as output devices. The whole system was centrally controlled, with users punching buttons on their telephone keypads in order to navigate through menus of preprepared information. Billions of dollars were spent on videotext experiments, but none of them included ways for the medium's users to communicate with—much less create content for—each other.

There was a sense of something in the air when the microprocessor was invented in 1971, but the acceleration rate of the cultural change to come wasn't visible yet to nonengineers like me. The first home computer kit, the fabled Altair, wouldn't be available until 1975 (inspiring Bill Gates to drop out of Harvard to write software for it). It's hard to convey to people who didn't live through it how impossible the first decades of the PC would have seemed if a time traveler had tried to tell us what was ahead. Ordinary humans did not take front-page news photographs with phones they carried in their pockets, or make their own brand of electronic entertainment and send it out on their own accord to people all over the world. That's what big newspapers and television networks, book publishers, and record labels were for.

In 1974, I came across Ted Nelson's self-published book *Computer Lib*, a talismanic object for the *Whole Earth Catalog* predecessors of cyberculture; like the *Whole Earth Catalog*, which had been published six years prior, in 1968, *Computer Lib* was oversize, full of illustrations, sidebars, and non-linear text, and looked like it was pasted together on a kitchen table.[42] Nelson foresaw a future of personal empowerment as soon as everybody could afford to own a microcomputer; he also envisioned a vast network of documents and media, all connected with hyperlinks. I didn't interview Nelson until a decade after I stumbled on his book, when the revolution he foretold was well under way. That's the aspect of exponential growth that can sneak up on you—progress in the 1980s was much faster than the development of personal computing in the 1970s, and by the late 1990s, computing devices in toys were becoming literally billions of times more powerful than the Department of Defense behemoths of the 1960s.

In 1974, a company called the New York Information Bank also appeared on my radar (which in those days consisted mostly of trips to the library and telephone calls to sources).[43] I went to the Information Bank's office—a two-room suite in one of the first high-rise apartment-office buildings in San Francisco. I wasn't one of the institutional customers the company was seeking, but the man who ran the office found my enthusiasm convincing. He directed me to a desk, where I was able to place one of those big Ma Bell landline handsets that could be used as a hammer if necessary into a rubber coupling device atop a large box full of, presumably, electronics. My modem sent a coded series of beeps and boops to a computer in New York that was listening on a long-distance phone line, and that responded with the characteristic shrieks, static, and electronic whooping noises that old-time modem users will recall. What I gained for all that work was access to a slightly amplified card catalog. I could find article references and their summaries. The speed of data transmission was around 110 bits per second, so downloading even a relatively short ten- or twenty-page article took long enough for me to go out for coffee while I was waiting. But I was able to print out references on whatever topic I was pursuing at that time as a freelance writer, and then bring the printout with me to the library. I felt like a man from tomorrow when I thumbed the paper card catalog at the public library, printout in hand.

In the late 1970s, two small companies—Apple and Microsoft—and their new category—PCs—began to attract press attention, along with scores of enterprises that few remember (I recall wondering whether to get the Apple II or Exidy Sorcerer, for example). I went to one of the first PC conferences, the legendary Computer Faire, convened in San Francisco's Civic Center by founder Jim Warren, who rode around the vast convention hall floor on roller skates, weaving through the hundreds of nerds (before the word was vaguely complimentary). I understood little of what was going on, and was not that strongly attracted to it. There wasn't much you could do with the first PCs except play games and program in the BASIC language—Gates's first product. But I picked up a flyer about using PCs to write with, and that *did* attract my interest.

In the mid-1970s I was using the state-of-the-art correcting electric typewriter. By pushing the right button, I could make my typewriter magically type backward over the last line I had entered with a white ribbon that overwrote my previous typing. The notion that I could move my cursor around and manipulate entire blocks of text was extremely appealing in a brute-force labor way. At that time, I typed out a page of draft, corrected it with pen, and sometimes physically cut and retaped different parts of the page. Then I had to retype the page. To me, not retyping a page was enough in itself to pursue my investigation of the flyer I had picked up. I was far too unschooled to understand much of what I read in the enthusiasts' publications, and nobody else was interested in paying me to do a story on using PCs to write with. Yet I drove from my home in San Francisco to Cupertino, about forty-five minutes away, to talk with a fellow named Jef Raskin, an employee of an Apple Computer Company, which was still small enough to occupy two buildings. Raskin later initiated the project that became the first consumer PC with a point-and-click interface: the Macintosh.

Indeed, Raskin had written his own program for using one of Apple's first PCs as what I learned to call a "text editor." Computer programmers knew about screen-based text editors because that was what they used in the post-punch-card days to edit programs on television-like screens before submitting them to the computer. Unfortunately, as Raskin told me, the visionary founders of Apple were convinced that their users would mostly use the machine to play games and program in BASIC. So the hardware only supported uppercase letters.

Toward the end of 1977, one of the magazines I scanned (in the old-fashioned sense) at the public library, *Scientific American*, published an article by Alan Kay titled "Microelectronics and the Personal Computer." When I came across the article a couple years after it was published, the first paragraphs of Kay's piece jumped out at me. As Kay famously noted, "The best way to predict the future is to invent it," which apparently he and his colleagues were doing at a facility within an hour's drive of my office.[44]

The article included photographs of a place where people actually moved paragraphs around by pointing to them on a screen, using a device called a mouse. The Xerox Palo Alto Research Center (PARC), designed in a style I later characterized as "Aztec modern," seemed like Wonderland, Atlantis, and Shangri-la rolled into one. Kay zoomed me back to a much-larger vision than the current popular fascination with the boy wonders and their jillion dollar start-ups—a vision of personal digital media as tools for powerful new means of creating, communicating, teaching, and learning. "The future increase in capacity and decrease in cost of microelectronic devices will not only give rise to compact and powerful hardware but also bring qualitative changes in the way human beings and computers interact," Kay observed. "In the 1980's both adults and children will be able to have as a personal possession a computer about the size of a large notebook with the power to handle virtually all their information-related needs."[45]

Although I had been unaware of it, Kay and others had been working on a highly visual, networked PC system since the early 1970s. By that time I was juggling jobs as a freelance writer. Before I got my hands on a point-and-click computer, one of my freelance jobs was as a part-time staff writer at the Institute of Noetic Sciences, a think tank devoted to the study of consciousness. It was at IONS, as we called it, that I started using a primitive screen-based PC program known as Wordstar. I initiated a campaign to land a writing job at PARC. I found the telephone number of PARC's public affairs director. She has since passed away, but I kept in touch with her for years because she gave me an important break. I called her every Friday and asked if she had any freelance writing work. On the third or fourth Friday, she said that they needed someone to work all weekend on scripting a slide show for a new product demonstration. After that, she started hiring me to help their wizards compose something about the technologies they were too busy inventing to write about.

It didn't take me long to find my way to Bob Taylor, who was then still the director of PARC's Computer Systems Laboratory, where the legendary Alto (acknowledged to be the first PC), the Ethernet network, and the laser printer had all been invented, and the graphical user interface was developed, extending Engelbart's ideas. I drove for forty minutes each way from my home in San Francisco to PARC in order to use an Alto to write as well as talk with people like Taylor. And they paid me for it. It was heaven.

Taylor had been a twenty-six-year-old research director at the Department of Defense's Advanced Research Project Agency (ARPA) in the 1960s, when interactive computing (meaning a programmer could enter commands and receive output from computers without submitting decks of punched cards to operators), computer graphics (an outgrowth of the air defense system), and Engelbart's Augmentation Research Center were just getting off the ground. In the Vietnam War era, when Congress forced ARPA to crack down on research that wasn't directly related to weaponry, Taylor recruited all the young talent he had previously funded for ARPA to join a new research laboratory that Xerox Corporation was starting in California. C. Peter McColough, the visionary Xerox CEO at that time, bankrolled a research center that would turn his company from a copier manufacturer to "the architect of information" for the office; infamously, the company wasn't able to seize the advantage from the market it had invented before Apple and Microsoft stole its thunder.[46]

By 1984, when the Macintosh launched, Taylor was tired of hearing about Steve Jobs and Gates. Apple and Microsoft had created toys compared to the handcrafted workstations at PARC, and the expensive office versions Xerox was trying and failing to sell. Taylor wanted to talk about even more interesting people than the teenage millionaires in the news—people who weren't featured in national magazines but who had made PCs possible, such as Engelbart. At Taylor's suggestion, I read Engelbart's 1962 paper "Augmenting Human Intellect," and was electrified by it.[47] In this paper, twenty years old by the time I read it, Engelbart detailed exactly how and why a computer could be used as a mind amplifier. More excitingly, Taylor told me that Engelbart had built his dream machine—Taylor had funded it when he was at ARPA—and was still actively developing his original vision. I made a pilgrimage to Engelbart's Augmentation Research Center, which had been sold by Stanford Research Institute after ARPA monies dried up, to Tymshare, a company that no longer exists. Engelbart's office was ironically in a building surrounded by the growing Apple campus. I drove down to interview him—an encounter that changed my life.

Engelbart couldn't help trying to recruit others to assist him realizing the idea that came into his head in the 1950s as he drove to work in the fruit orchards that were to become Silicon Valley. Making it happen turned out to be more difficult than he had imagined. After a decade of trying to convince computer scientists and the computer industry that their technology could amplify human cognitive abilities, Engelbart wrote his paper because he realized that nobody even had a conceptual framework or mental model of computers that would enable them to see their potential. He certainly succeeded in convincing me of his vision, just as he had previously attracted engineers to build his first prototypes and had changed the way computer designers thought about what they were doing when he pulled off the famous "mother of all demos" in 1968: at an assembly gathering most of the computer designers in the world, he showed off the point-and-click hypermedia system his Augmentation Research Center had developed.[48] I remember driving home from my first meeting with Engelbart, all fired up to write about someone who had offered an idea that was changing the world, and had done so out of a conviction that he had a duty to use his knowledge to facilitate ways for people to work together to solve the world's ever more complex problems.

One aspect of Engelbart's vision, though, hasn't quite yet come to pass in the way that his first prototypes of graphical user interfaces, hypertext, multimedia, and online knowledge communities have developed into global media in his lifetime. In his original paper, Engelbart described a system of "humans using language, artifacts, methodology, and training."[49] I recall Engelbart remarking to me on several occasions recently that the artifacts' development had far outstripped the cognitive and social aspects of an augmentation system as he saw it—the language, methodology, and training had not spread through the population the way home computers with mice, icons, and hyperlinked networks had caught on. The books I have written since I met Engelbart, culminating in this one, have attempted, in my own humble way, to help remedy that deficit.

Reading Kay, and meeting Taylor and Engelbart, led me to write a book about "the history and future of mind-amplifying technology," as my *Tools for Thought* was subtitled. In the process of researching that book, I bought a twelve hundred bit-per-second modem (today's broadband speeds are millions of times faster) for five hundred dollars in 1983, started exploring amateur computer BBSs, and joined the Whole Earth 'Lectronic Link (WELL) when it was a few months old, in summer 1985. Two years later, I wrote my article on virtual communities for *Whole Earth Review* that apparently put the term into the public vocabulary, as noted earlier.[50]

I started writing about my life online in part because the small subculture of enthusiasts for computer-mediated communication that existed in 1985 was certain that what we were doing would become important in the future, and in part to justify to my wife all the fun I was having hanging out online. I admit that I was and remain an enthusiast for social media of all kinds. I maintain and participate to this day in blogs, vlogs, wikis, social network services, and BBSs. My outlook, however, has grown more critical over the years. The online culture has changed. I've changed. And how I think about the significance of online socializing has changed. While still a devotee, I'm now aware and wary of the rat holes, hidden biases, unwholesome interchanges, and delusions of grandeur that can plague online culture. It is possible, I have long believed, to temper one's ardor with critical thinking, and that it is not healthy to have to choose only between being a complete supporter and a total skeptic. I want the reader to keep in mind that the advice I'm giving about how to participate productively in digital culture grows out of my enthusiastic, if not uncritical, use of these media.

I admit that I'm immersed. I understand that this immersion works for me in my particular situation, sitting in my garden as I tap out these words under the plum tree, in ways that it doesn't work for many others, and I believe that this rate of mediated communication should be regarded with a critical eye for multiple reasons. Yet compared to thirty years ago, in my typewriter and library stacks days, I guarantee you that as a thinker, writer, learner, and teacher, both my ability to know and communicate have been immensely empowered—from the self-correcting typewriter to the iPhone, from the local library to all the knowledge in the world floating in the air, from the card catalog to Google. Now, if I can only figure out how to stay off-line for a few weeks without having to deal with ten thousand messages, how to better detect texting drivers, or the best strategy for trying to teach thirty students while they surf the Web.

As laptop-carrying, smart-phone-using members of the digitally connected infosphere, we need to start by learning a new discipline: the literacy of attention. As citizens and cocreators of the cultures that shape us, we need participatory media skills. As collaborators in the collective intelligence that faces massive problems from global warming to water-sharing conflicts, we need to learn literacies of cooperation, mass collaboration, and collective action. As dwellers in the network society, we must understand and master the nature along with use of social networks, technical and human—and grasp the way both mediated and face-to-face social practices can increase or drain social capital. And in a world where nobody can trust the authority of any text they find online, the ability to quickly evaluate the validity or bogosity of information is no longer an intellectual nicety. Critical thinking about media practices has become an essential, learnable mental skill.

My attention—the symbols, sounds, and images I personally experience—is the thread that weaves these dimensions into an integral whole. What use to me are fiber optics and network protocols without my attention as well as thought processes to make sense of all the bits flying around the networks? Attention connects the events that occur simultaneously in the mind, between people, and among technologies. Human thought processes are themselves no more than a part—a kind of focusing lens—of a system that includes neurons, symbols, search engines, social systems, and computational clouds.

1 Attention! Why and How to Control Your Mind's Most Powerful Instrument

In the transmission of knowledge the children and teachers of the future should not be faced with a choice between books and screens, between newspapers and capsuled versions of the news on the Internet, or between print and other media. Our transition generation has an opportunity, if we seize it, to pause and use our most reflective capacities, to use everything at our disposal to prepare for the formation of what will come next. The analytical, inferential, perspective-taking reading brain with all its capacity for human consciousness, and the nimble, multifunctional, multimodal, information-integrative capacities of a digital mind-set do not need to inhabit exclusive realms. Many of our children learn to code-switch between two or more oral languages, and we can teach them also to switch between different presentations of written language and different modes of analysis. Perhaps, like the memorable image captured in 600 BCE of a Sumerian scribe patiently transcribing cuneiform beside an Akkadian scribe, we will be able to preserve the capacities of two systems and appreciate why both are precious.

—Maryanne Wolf, *Proust and the Squid*, 2007

I've taken to opening the first class session of each semester with words that always seem to capture students' attention: "Close your laptops." My next words, "Turn off your phones," come as no surprise at that point. Then I tell them to shut their eyes, which does seem to startle them. I ask them to take one minute to observe how attention leaps effortlessly from thought to thought, directing them to "note how you don't have to work to make your mind wander. It does that all on its own."

By 2005, the year Facebook spread to most universities, the sight of students staring at their laptops rather than looking at their professor had become commonplace in classrooms. Nevertheless, the spectacle puzzled me the first time I witnessed it. Were they taking notes? Discussing my lecture? Messaging each other about what to eat for dinner? Watching YouTube? After six years of observing college students, asking them directly, and even recording their actions with video, I've concluded that the answer is "all of the above." The attentional shift that has been taking place among students for some time is now propagating far beyond the campus: all people and media are available all the time, and in all places, but relatively few people appear to use ubiquitous informational access and social connectivity politely and productively. Students and professors were among the first but not only humans whose social norms have been disrupted by the use of always-on/everywhere media.

"I can't change your mental habits in a single semester," I continue, after they open their eyes, "but I can suggest a simple, powerful idea: you can learn to be aware of how you shift your attention when your phone buzzes or your laptop screen beckons." I've found that introducing a little mindfulness where previously there had been none can be insidiously irrevocable. Asking students to become conscious of their laptop use during class is like asking them to not think of a purple dinosaur.

Each week, I introduce a new attention probe to the classroom. I told a cohort of fifty students, for instance, that five of them could have their laptops open at any one time. "In order for somebody else to open their computer," I stipulated, "one of the current five will have to close theirs." This was not only an attention probe but also a collective action problem. It forced the current five to be aware of their own attention in the context of other students who were waiting to Google my lecture (or slay monsters in a role-playing game). Each class session, I reminded students that the objective was "to get you to start paying attention to the way you pay attention."

I realized that my students had no idea of what it felt like to stand in front of them while they were concentrating on their computer screens. With the students' permission, I made a video of how my classroom appeared from where I stood and then projected it at the front of the room.[1] I also aimed another camera at their laptops from the back. As I suspected, I captured a visual record of one student's attention meandering. First, he started inspecting my personal Web site. Then he checked his email. He moved between these separate Web sites and cognitive tasks in less time than it has taken you to read these last few sentences. I wasn't surprised to discover that students surf the Web during my class. What did intrigue me, though, was that this particular student was in fact one of my *most* attentive students. If I were to have asked him a question while he was checking his email, I'm certain that he'd know the right answer. Was this young man born with the talent to juggle multiple parallel information streams without dropping anything, the way some people are born to run swiftly? Or more interesting, Could it be that he was no more or less attentionally endowed than other students but had *learned* something that others could learn?

I posed the last question to Clifford Nass, the often-quoted social psychologist who has an office two doors from mine at Stanford University. Nass was the principal author of a widely cited study that showed most media multitaskers to be worse than they thought at multitasking. His team also reported that their subjects' performance on individual tasks degraded significantly when they attempted to multitask.[2] The most important thing I learned from Nass about mediated attention is that most media multitasking is actually task switching rapidly, *not* parallel processing, and this switching is more mentally costly than anybody thinks. People multitask because they believe they can get more done, but Nass has solid evidence that the opposite is usually the case. I meditated on this at some length, since I frequently live with multiple, simultaneous attention streams myself. Many popular publications have interpreted Nass and his colleagues' research to mean that multitasking doesn't work, period. I was intrigued, however, by the few exceptions to Nass's and others' findings about the inefficiency of task switching—the evident existence of a small number of *successful* multitaskers.

Researchers at the University of Utah reported that about 2.4 percent of the college students in their studies were able to talk on the telephone while operating an automobile driving simulator—*without* degraded performance.[3] At the time of this writing, nobody knows whether those "supertaskers," like the student I captured on video or those at the University of Utah, are innately better at processing multiple streams of information or stumbled on mental tricks that enable them to outperform others.[4] I do know of one instance where the ability to switch rapidly and without loss of ability from task to task is essential: aviation. Indeed, one recent study claimed that experienced fighter pilots handled their executive control functions more effectively than a control group of similar intelligence: "The pilots displayed superior cognitive control, showing significantly greater accuracy on one of the cognitive tasks, despite being more sensitive to irrelevant, distracting information."[5] Fighter pilots have to hyperfocus on the current target in their sights ("superior cognitive control") without losing awareness of the speck in their field of vision that might become a threat in a few seconds ("irrelevant, distracting information"). When I posted a link to this research on Twitter, someone I had not previously known replied within seconds that aviators must simultaneously "aviate, navigate, and communicate." I reviewed the scientific literature about attention training and found ample clues to actual techniques for improving one's ability to pay attention in the context of multiple competing media.

The first thing I learned is good news if you have been thinking that "attention training" sounds like too much work: you can experience immediate benefits by beginning in small ways to exercise mindfulness regarding your attention online. In this realm, taking *some* control, even if it is a baby step, is far better than passively letting your attention be grabbed without reflection. Growing evidence indicates consistent exercise can strengthen self-control of attention.[6]

I'll help us begin the process of learning to control attention by examining how attention works. Then I'll consider the dangers of distraction posed by social media, examine arguments that the Web is making us stupid, and weigh counterarguments that we can learn to use digital media to multiply intelligence. I'll scrutinize multitasking—how it works, why and when it doesn't along with what it's good and not good for. Learning the latest knowledge about the brain's capacity to rewire itself—known as "neuroplasticity"—can increase your power to actually direct your brain's self-rewiring function rather than just being affected by it.

It is not possible to explain the cognitive underpinnings of attention in a simple way without oversimplifying, so keep in mind that my objective is not to convey a rigorous review of brain science but instead to transmit enough basic knowledge to enable you to wrest control of your attention from media that might be trying to capture it. Oversimplification number one: *attention, memory,* and *executive control* are the fundamental components of thinking—and the executive control process is the particular power you can tap to control your use of social media. To experience how these three components work together, reflect on the mental actions you probably undertook the last time you misplaced your keys. First you tried to recall where you last saw your keys. You moved various snapshots of memory from the "back of your mind" into the spotlight of your attention. Part of your brain coordinated attention and memory in a mental simulation of a spatial search.

In your mental quest for lost keys, your brain shifted remembered perceptions from the background to the forefront of your attention—into what cognitive scientists call "working memory." Working memory is well known among psychologists because of an intriguingly titled research paper published in 1956, George Miller's "The Magical Number Seven, Plus or Minus Two."[7] Miller demonstrated that most people can keep around seven chunks of information in their working memory (plus or minus two) at any one time. In order to work with more than seven chunks, some brain mechanism has to swap out the current information under attention's spot-light and fetch another chunk from memory.

The part of your brain that you use to retrieve memories and keep information in your working memory is referred to as the "executive control" or "cognitive control" function. Scientists often liken executive control to a kind of CEO of cognitive processes that coordinates attention and memory. I caution against investing too much in the image of a little person in your brain who acts in this manner. Executive control is a function, not a person. (If you posit a little person inside your brain, you still have the problem of describing how that little person makes decisions. Do they have yet another little person nested within them?) I think of executive control as akin to a self-conducting orchestra. The executive function also helps us think about our thinking—a process technically known as "metacognition," the brain function we can use to gain control of our lives online.

Like a conductor pointing at a soloist, executive control chooses from moment to moment which memory, thought, or perception to foreground, and which ones to recede into the background, based on a previously decided goal (like the conductor's musical score). The central executive function must terminate all active selection of memories, thoughts, and perceptions related to a prior goal in order to activate the cognitive processes associated with a different aim. Suppose, for example, that your baby is crawling across the floor while you are trying to remember your keys. Keeping your baby alive is an overriding priority (strategic goal), so it is necessary to suspend your mental search for your keys (previous strategic goal) while you look for dangerous objects in the baby's path. This rerouting of attention is what your brain does when you think you are multitasking. Your brain expends time and energy whenever it suspends one attentional process to fire up another one.

When you shift your attention, there is always a short interval during which you must reorient, refocus, and filter out competing information in order to move from one stable theme to another, whether you move from remembering your keys to saving your baby, or from working on a report to reading your email. Cognitive scientists call this temporary disruption the *attentional blink*.[8] Only you can tell whether shifting attention outside the boundaries of your focal theme is worth the time lost getting back to your original task ("switching costs")—but given experimental evidence, most people have to face the fact there *is* a cost, most of the time. Gloria Mark, professor of informatics at the University of California at Irvine, has studied the effects of workplace interruptions and found that it can take up to a half hour to regain concentration on a central task afterward.[9]

When you scan the environment for objects of interest like crawling babies, your attention is widely focused—like turning on the lights in a room to take it all in at a glance. When you notice the fork in the hand of your crawling baby and estimate that she is five seconds away from jamming it into an electric outlet, your focus narrows to a spotlight. Everybody uses each mode all the time—otherwise nobody would be able to walk through traffic, nor would anyone be able to perform brain surgery. We are at least minimally aware of the broad margins of our perceptions even while we are narrowly concentrating most of our attention. Our senses are receiving an estimated eleven million bits of information per second, while we are conscious of only forty.[10] When talking about attention, mostly we are referring to the forty bits that are under the tightly focused beam of attention—although research has demonstrated that the remaining millions of bits are being attended to on a subconscious level, and still affect our thoughts and behaviors.

Humans pay a lot of attention to other humans—hence the success and seductive distractions of social media such as Facebook and Twitter. The discovery of "mirror neurons" in primates strongly implies that paying attention to others is one of the few human cognitive capabilities that may be neurally "hardwired." Mirror neurons fire when you do something, but they also fire in the same way when you watch someone else doing the same thing. The scientists who discovered mirror neurons believe they are fundamental to social behavior: "If we want to survive, we must understand the actions of others. Furthermore, without action understanding, social organization is impossible. In the case of humans, there is another faculty that depends on the observation of others' actions: imitation learning. Unlike most species, we are able to learn by imitation, and this faculty is at the basis of human culture."[11]

Neuroscientist and reading specialist Stanislas Dehaene also believes the trainability of our attention to others is essential to human sociality.

In the human species, cultural selection is further amplified by its intentional character. As stressed by the primatologist David Pratt Mac, Homo sapiens is the only primate with a sense of pedagogy. Only humans attend to the knowledge and mental states of others in order to teach them. Not only do we actively transmit the cultural objects we find most useful, but—as is particularly apparent with writing— we intentionally perfect them. More than 5000 years ago, the first scribes hit upon an extraordinary potential deeply embedded in our brain circuits: the possibility of conveying language through vision. This initial idea was then perfected by generations of scribes. A long chain of teaching tradition links us to these early writers who worked diligently, from one generation to the next, to make their invention easier to assimilate by our primate visual system.[12]

Mirror neurons might be basic to introspection as well as social behavior and learning. Running our own mental simulations of the future, the attentional process we will later learn to recognize as the "narrative network," turns our ability to mentally rehearse something we've seen others do into a tool for controlling our own awareness.

When it comes to interacting with the world of always-on info, the fundamental skill, on which other essential skills depend, is the ability to deal with distraction without filtering out opportunity.

Conscious Distraction: Are You Captain or Captive of Your Attention Muscles?

In regard to propaganda the early advocates of universal literacy and a free press envisaged only two possibilities: the propaganda might be true, or it might be false. They did not foresee what in fact has happened, above all in our Western capitalist democracies—the development of a vast mass communications industry, concerned in the main neither with the true nor the false, but with the unreal, the more or less totally irrelevant. In a word, they failed to take into account man's almost infinite appetite for distractions.
—Aldous Huxley, *Brave New World Revisited*, 1958

While I was writing this book, my friend Duke University professor Cathy Davidson was also working on her own book about attention.[13] In her blog, Davidson recounts an incident that happened when she was tracking down footnote references requested by the editor. She was working at her desk, got up to put a teakettle on the stove, and went back to her writing. Hearing a garbage truck outside, she assumed it was the source of the burning rubber she was beginning to smell. When she started to see smoke, Davidson realized that the water had boiled out of the teakettle and the plastic handle had been melting. She had forgotten to pay attention to the stove while concentrating on her book. So she consciously told herself to be particularly mindful when she got into her car a few hours later, reminding herself that her thoughts were likely to drift back to those footnotes. Davidson willed herself to be more vigilant than usual while driving. A few minutes later, two dogs darted in front of her car, and she hit her brakes on time.[14]

Davidson told this story to emphasize the need to exercise conscious decision making in order to avoid the burning teakettle, dogs in the road, or online distraction:

The simple-minded accounts of how the Internet is ruining our attention would not connect the teakettle with the stray dog. But the new neuroscience of attention says that itemizing attention is wrong. Interconnection is all. *Because* the trauma of the fiery red kettle caught my attention and almost got me in trouble, I was able to recalibrate, very consciously and self-consciously, my attention level and make adjustments. I'm not sure I would have done so before starting this book because we tend to think of attention passively, as if it is out of our control or only controlled outside of ourselves, by the world around us. That is not true. We can track our own attentional pathways and learn from them.[15]

When you are online, how often do you control your own focus—and how frequently do you allow it to be captured by peripheral stimuli? At least some of the time, your attention *is* controlled by outside forces. Jumping at a loud noise or applying the brakes at the sight of a dog in the road, for example, is not the product of reflection. Willed, sustained attention, however, requires premeditation. With a goal in mind, I actively maintain attention on a specific objective such as writing this sentence and simultaneously filter out information that is not directly applicable to my central task, such as the tiny icon that popped up in the corner of my computer screen to alert me to new email. This skill at screening out information before it reaches full awareness is not something that social media itself can do for me; automated filters help, but the most important filter is a function of my brain, not my PC. Only *you* can know your goals, and only you can determine which stimuli are relevant at any moment.

The executive control we all exercise when we maintain focus on one task becomes useful when we move from understanding attention to controlling it. Cognitive psychologists have studied how attention sharpens the field of concentration in part by filtering out unwanted stimuli.[16] In a crowded setting, you employ executive control of your attention when you listen to one voice and tune out the others around you (this classic example is known, appropriately, as "the cocktail party effect").[17] If you want to experience your use of executive control directly, simply switch your attention between different conversations at a party. Observe how you change filters without consciously sorting through the competing information streams—suddenly, the voice you were following attentively becomes one of those you are *not* following. What do you move when you shift your attention? What did Davidson adjust when she jumped up from her desk and rescued her melting teakettle before it burst into flames—or when she decided to be hypervigilant when she started driving her car that morning?

Gaining control of your attention while you are online requires, first of all, intention. When you formulate a goal, you need to *intend* to achieve it. Goals and intentions enable your executive control to attune to the part of your information environment that matters most, and tune out what is irrelevant, at least for the purpose of your goal. Even when you think you are focusing effectively, you might also be blocking out significant information—perhaps not relevant to your immediate task but instead vital to other strategic goals—along with all the distractions you are better off ignoring.

This phenomenon, known as "selective inattention," is dramatically illustrated by the online video of the "awareness test" conducted by Daniel Simons of the University of Illinois at Urbana-Champaign and Christopher Chabris of Harvard University.[18] Subjects were asked to watch a short video of two groups (distinguished by black or white T-shirts) passing basketballs and count passes by one team, or keep track of bounce versus aerial passes. While the basketballs were passed, an actor walked through the scene wearing a gorilla suit, paused, turned to look at the camera, and walked on. When asked whether anything out of the ordinary occurred, around 50 percent of the subjects did not report seeing the gorilla. The assigned task created a frame for the subjects' attention, filtering out distractions that didn't fit, to the point where a gorilla on a basketball court escaped notice.

When external stimuli tempt attention away from the intended focal point, the external stimuli are usually regarded as distractions—a word that has a time-wasting connotation. Distraction sounds sinful (sloth, probably). *Distracted: The Erosion of Attention and the Coming Dark Age* is the title of a contemporary critique of always-on consciousness (which I will consider shortly).[19] I'll start with the general dangers of distraction before considering several contemporary arguments that the way we use the Internet is, as Carr puts it, "making us stupid."[20] Yet I want to set a boundary on our cautions: the war on distraction can go too far for your own good. Distraction is a real issue, but dwelling exclusively on its dark side can be a form of selective inattention. Alison Gopnik, professor of psychology at the University of California at Berkeley, warns that

it is certainly true that by the time we're adults attention is a limited resource and attentional patterns are hard to change. But the exaggerated highly-focused attention we consider appropriate in a contemporary classroom is itself a recent cultural invention, and one with costs as well as benefits. Guatemalan Mayan mothers successfully teach their children to divide their attention, as Western mothers teach children to focus theirs.[21]

Without the capacity for distraction, you wouldn't hear the taxicab horn when you step off the curb (perhaps while you are concentrating on your BlackBerry). You wouldn't want to train yourself to ignore distractions when your life might be at stake. If you want to make mindful use of media, however, you'll need to train yourself to recognize and withdraw attention from activities unrelated to your intended goal of the moment.

Media-triggered distraction can be:

- *Unproductive* for the goal oriented
- *Unhealthy* for everybody
- *Fatal* for a growing number
- *Addictive* for some
- An invitation to *bad parenting*
- *Socially alienating*
- A cause for a dangerous loss of solitude

The meaning of unproductive, like distraction, requires both context and a firm idea of one's goals. If your aim is to produce a certain amount of external output (as opposed to the more internal production of learning), then the invitations to serendipity, play, and digression that digital media offer are obstacles and dangers. If your aspiration is to learn, help build community, and explore, then the issue gets more complicated. I'll revisit this when I talk about strategic goal setting.

Although she isn't a cognitive psychologist or neuroscientist, Linda Stone was an obvious choice when I started inquiring into the connection between attention, always-on media, and health. Stone has in fact been immersed in creating online media for the twenty-five years I've known her. I first met her when she was one of Apple's multimedia researchers in the 1980s. In the 1990s, when she was director of Microsoft's Virtual Worlds Group, Stone and I sat in my garden to discuss virtual communities. Since she retired from Microsoft, Stone has been concerned about the ways social media use might be affecting our minds and bodies. She was kind enough to make another garden visit this past summer to converse about our mutual interest in literacies of attention.

As we sat under my plum tree, Stone recalled that she had noticed something crucial about her own online behavior while she sat at her computer one day. "I realized that I hold my breath sometimes when I am doing my email." She has recounted this little epiphany in print:

I've just opened my email and there's nothing out of the ordinary there. It's the usual daily flood of schedule, project, travel, information, and junk mail. Then I notice . . . I'm holding my breath. As the email spills onto my screen, as my mind races with thoughts of what I'll answer first, what can wait, who I should call, what should have been done two days ago; I've stopped the steady breathing I was doing only moments earlier in a morning meditation and now, I'm holding my breath.[22]

Stone grew even more intensely interested when others reported that they, too, sometimes held their breath while reading or writing email—a phenomenon that she started calling "email apnea." She told me that she came to realize that "breathing is the regulator of attention." Stone reminded me that holding one's breath is directly connected to the "fight or flight" response. When your ancestors and mine heard a noise, they held their breath until deciding whether to flee, fight, or ignore the sound, while their glands pumped energy-mobilizing hormones into their bloodstreams, just in case. Holding your breath affects the body's balance of oxygen, carbon dioxide, and nitrogen oxide. It activates the sympathetic nervous system, causing an increase in glucose and cholesterol levels in the bloodstream along with an increased heart rate as well as a sense of hunger. Stone remarked that regular breathing patterns, by contrast, activate the parasympathetic nervous system, causing relaxation, the release of digestive enzymes, and a sense of satiety—signs of a "rest and digest" mode. She pointed out that "we're putting our bodies in a state of almost constant low-level fight-or-flight. This is great when we're being chased by tigers. But how many of those 500 emails a day is a TIGER? How many are flies? Is everything an emergency? Our way of using the current set of technologies would have us believe it is."[23] Paying attention to your breath—the core technique of mindfulness meditation methods—is where Stone suggests starting to moderate our online reactions. I'll get back to that later. For now, I'm convinced that Stone is right to think that attention to breathing could be a tool to help moderate our unthinking, ultimately unhealthy reactions to many online stimuli.

In one arena of daily life, distraction has proven to be life threatening. Who hasn't witnessed the chilling sight of another driver in the next highway lane who appears to be texting while driving? A Harvard study in 2003 estimates that 2,600 traffic deaths and 330,000 accidents annually are caused by cell phone distractions.[24] A study in 2009 of professional long-haul truck drivers who equipped their cabs with video cameras for eighteen months claims that the collision risk became twenty-three times greater when the drivers texted.[25] University of Utah researchers found that drivers who talked on a cell phone—just talked, not texted—were as impaired in driving simulation tests as subjects with blood-alcohol levels close to the legal limit.[26] Although there are more subtle dangers to consider in this chapter, texting while driving kills; that's all that needs to be said about it. I'll only add that the fact that anyone would risk life and limb for an LOL is a clue that something about texting hooks into the human propensity to repeat pleasurable behaviors to the point of compulsion.

Is the compulsion to check up frequently on our online connections an addiction problem? I want to be careful when using the word *addiction*, which is also used for serious physiological dependencies. Nevertheless, some aspects of social media behavior that many of us experience bear an uncomfortable resemblance to graver compulsions. In the *Harvard Business Review*, Tony Schwartz proposes that information overload is a phenomenon we bring on ourselves as we become habituated to regular doses of information, social networking, and messages. Schwartz recounts a talk he gave about the value of doing one thing at a time, and how he removed the incoming streams of interruptions from email, instant messaging, and text messages. He was struck when a young man working in finance stated, "I believe everything you said, but I can't do it. If I get an email, I have to look at it." Schwartz responded by asking: "Have you considered just turning it off at certain times during the day?" "I don't think I can," the young man replied. "As soon as I turn it off, I'd start obsessing about what I'm missing."[27] Do you know how the young man feels? I do. My twenty-six-year-old daughter does.

That doesn't mean we have to surrender to obsession. And it doesn't necessarily mean that all task switching is detrimental to one's focus. Recent research reported in the journal *Cognition* offers evidence that brief distractions from a focal task may improve concentration over the longer run: "We propose that deactivating and reactivating your goals allows you to stay focused," the study's authors said. "From a practical standpoint, our research suggests that, when faced with long tasks (such as studying before a final exam or doing your taxes), it is best to impose brief breaks on yourself. Brief mental breaks will actually help you stay focused on your task!"[28]

The craving for digital stimulation may be similar to food and sex addictions, which each arise when healthy behavior warps into a compulsion, eventually impairing the individual's functioning.[29] How do we get addicted to social media? Emily Salvaterra and her colleagues propose a psychophysiological process that reinforces Internet use, and then escalates into obsessive-compulsive "checking behavior."[30] They offer Youmasu J. Siewe's outline of addiction and its behavioral development: first, indulgence in the addictive behavior or substance produces pleasure, which leads to a craving for more, which if indulged too regularly leads to withdrawal symptoms, loss of control over the addictive behavior, the need for more and more frequent as well as larger doses of the addictive substance or behavior, which ultimately produces negative consequences for the addict.[31]

This book is not for people who have a serious problem controlling their online behavior. Still, some of the dynamics of addiction are probably at play in a weaker way for most of us who spend a large part of every day online. Nobody who has ever hit the "refresh" link on their email in-box or Twitter timeline every few minutes (or more often) can deny that the possibility of compulsion lurks behind the undeniable pleasures of social media. Matt Richtel in the *New York Times*, echoing Stone, proposes that urges to check social media to the point of compulsion "play to a primitive impulse to respond to immediate opportunities and threats. The stimulation provokes excitement—a dopamine squirt—that researchers say can be addictive. In its absence, people feel bored."[32] Any simple explanation of the way hormones regulate behavior is bound to be inadequate, and "dopamine squirt" seems to be a favorite phrase among those who fear that entire populations have become addicted to social media. A range of experiments has shown that the hormone dopamine does indeed appear to be associated with a reward for "seeking" behavior.

Hormones may not wholly control our behavior, but biochemical reflexes undoubtedly still influence us, even though humans have made the transition from hunting and gathering on the savanna, to hunting and gathering online. In addition to the stimulant dopamine, another chemical, oxytocin—a normally occurring human hormone that appears to facilitate bonding between friends, lovers, or parent and child—appears to come into play as well, especially when social media take up a regular part of one's life. Oxytocin was first recognized for its facilitating role in infant-parent bonding, and today it has been described more generally as "the human stimulant of empathy, generosity, trust, and more."[33]

I can remember my excitement anticipating the arrival of the daily snail mail in my early years as a writer—as I put my dreams into stamped, self-addressed envelopes, and months later, would receive a rejection letter or contract. The brain-body rush I got when I heard the sound of the mail slot downstairs certainly seems akin to the descriptions of oxytocin highs. That I was rewarded infrequently by contracts rather than rejections only made my anticipation that much stronger, through the kind of "intermittent reinforcement" that makes slot machines addictive. I also recall how thrilling it felt at first to see "you've got mail" notifications pop up on my computer screen—mail more than once a day!—and how I learned to interrupt my writing work to see who awaited in my mail queue. Intuitively, a hormonal component makes sense. In a single-subject case study, Paul J. Zak of Claremont Graduate University, a pioneer in the emergent neuroeconomics field, discovered that a spike in oxytocin occurred after using Twitter for ten minutes.[34] The implication of Zak's finding is that the frequent checking behavior of our favorite social networks might be reinforced by the chemically mediated feeling of connection that it gives us (oxytocin) as well as the chemical reward for hunting-seeking behavior (dopamine) described by others.

If you are a parent, beware of the media equivalent of secondhand smoke: the impact of how you use media in your children's presence. In an article about young people who "feel neglected by media-obsessed parents," Ellen Reagan told the story of a major advertising agency CEO who was, not surprisingly, a heavy BlackBerry user. One day, his daughter interrupted midtext by giving him two small, rolled-up pieces of paper. When he asked what they were, "sweetly, she slipped the construction-paper cylinders over his poised thumbs. "Blackberry handcuffs," she said."[35]

Daniel J. Siegel, clinical professor of psychiatry at the University of California at Los Angeles (whose work on mindfulness has been among my own key influences,) told Reagan that "children need their feelings to be felt by their caregivers. That is what creates secure attachment."[36] According to Siegel, paying attention to your kids is not just good parenting; parental distraction interferes with the emotional sense of attachment and can affect children's developing brains. In his book *The Mindful Brain*, Siegel makes a direct connection between parent-child attention and the individual's power to control attention:

It is the social circuits of the brain that we first used to understand the mind, the feelings and intentions and attitudes of others. When we view mindful awareness as a way of cultivating the mind's awareness of itself, it seems likely that it is harnessing aspects of the original neural mechanisms for being aware of other minds. As we become aware of our own intentions and attentional focus, we may be utilizing the very circuits of the brain that first created maps of the intentions and attention of others. . . . We can propose that the interpersonal attunement of secure attachment between parent and child is paralleled by the intrapersonal form of attunement of mindful awareness.[37]

Consider also the ways our adult media-compulsive behavior might interfere with our parental responsibility to teach adolescents how to control themselves. A *New York Times* article about a teenager in California who sent more than twenty-four thousand texts in a single month quoted the young woman talking about her mother's attempts to curb her texting habit: "She should understand a little better, because she's always on her iPhone."[38] Reagan also quotes MIT professor Sherry Turkle: "What is so poignant is that children try to bring their parents out of the 'BlackBerry zone' as they call it. Kids complain to me about parents having the phone with them as they watch sports with them, or TV movies, or go camping. One boy reflected on how his father kept the BlackBerry on the nightstand as he read him Harry Potter."[39] In a 2011 interview, Turkle told me about

stories of parents going to pick up their kids at school and the parents are not looking up from their BlackBerries. So the kid gets in the car and is absolutely crushed
because that's the moment when, even if that thirteen- or fourteen- or fifteen-year-
old pretends to be nonchalant, that is the moment when they want that eye contact.
They need that eye contact. Parents bring their kids to the museum. I sat there at the
children's museum for hours on end and watched those parents come in with their
kids. They let their kids go through the whole museum, and they're standing along
the walls scrolling on their BlackBerries and hitting their iPhones. They're losing the
point of the exercise, which is to be with your kid at the museum—because they're
happy that the kid is diverted and they're happy to be on their devices. That's why I
call the book *Alone Together*.[40]

When I consider social network literacies, I'll zoom in on the long debate
that sociologists have had about the effects of trains, telephones, or televisions on the quality of human social connection in large social groups, or
"society" in the aggregate. Sociologist Claude Fischer of the University of
California at Berkeley, author of the classic 1991 book *America Calling: A
Social History of the Telephone to 1940*, noted recently that "if you go back
100 years, people were writing things about the telephone not unlike what
people are writing about these technologies. There was a whole literature
of alarm—how it's turning everything upside down."[41] Whether or not our
broader social groups are growing more alienated as we grow more connected, critics I respect are voicing concern about the more atomic level of
interaction: relations among families and friends.

The book that Turkle mentioned above—*Alone Together: Why We Expect
More from Technology and Less from Each Other*—explores this danger to
interpersonal relations.[42] "We've come to confuse continual connectivity
with making real connections. We're 'always on' to everyone. When you
actually look more closely, in some ways we've lost the time for the conversations that count," Turkle told *USA Today*.[43] In my interview with her,
Turkle asserted:

Use technology as an opportunity to think about your values. Technology has been
a great gift. We have new possibilities for wonderful new things. But one of those
possibilities should not be sitting in the corner of the museum while your kid goes
through without you. I'm looking out over a park. It's filled with children and parents. The parents are on the bench not looking at the kids. We're sitting at dinner
and texting, while everybody in the family is not talking. What are we allowing the
technology to enable, and is it really where we want to be?[44]

I couldn't agree more that mindfulness about technology and family
discussions about it are necessary; I'm not convinced that the mediated
interactions many people engage in are necessarily the devil's work. We're

thinking and socializing differently, and I'm not entirely comfortable judging these changes from the values of the past.

Consider the possible danger of alienation from ourselves as well as others. What might be called "the Thoreau objection" to the siren call of digital distraction is worth examining, since the always-on availability of information to inform or amuse along with perpetual possibilities for social interaction may be depriving us of something humans have always drawn on: solitude. As Turkle told me in 2011, "If you don't know how to be alone, you will always be lonely. If you're always connected, from the age of eight, your default position is to only be connected and you don't learn the restorative virtues of solitude."[45]

I don't argue with the Thoreau objection. I embrace it. Years ago, I cut a door in my office wall; it's now three steps to my garden. The fact that I acknowledge my attraction to distraction doesn't mean that I have to succumb to the urge to be constantly connected. I simply ask myself when I reach for my iPhone while waiting in line, Why not stay disconnected for a minute and see what happens? Or I deliberately leave my podcasts at home when I take the dogs out for a walk in the neighborhood. Throw some sand into the machinery that automatizes your attention.

Distraction might be more than just deviation from focus. A good question for any mindful digital citizen to ask is, What are my media practices doing to my brain?

(Using) the Internet Makes Us Stupid (or Not)

When it came to writing, Theuth said, "This discipline, my King, will make the Egyptians wiser and will improve their memories: my invention is a recipe for both memory and wisdom." But the King said, "Theuth, my master of arts, to one man it is given to create the elements of an art, to another to judge the extent of harm and usefulness it will have for those who are going to employ it. And now, since you are father of written letters, your paternal goodwill has led you to pronounce the very opposite of what is their real power. The fact is that this invention will produce forgetfulness in the souls of those who have learned it. They will not need to exercise their memories, being able to rely on what is written, calling things to mind no longer from within themselves by their own unaided powers, but under the stimulus of external marks that are alien to themselves. So it's not a recipe for memory, but for reminding, that you have discovered. And as for wisdom, you're equipping your pupils with only a semblance of it, not with truth. Thanks to you and your invention, your pupils will be widely read without benefit of a teacher's instruction; in consequence, they'll entertain the delusion that they have wide knowledge, while they are, in fact, for the most part incapable of real judgment.

—Plato, *Phaedrus*, fifth century BCE

Recently, a heated public discussion was ignited by a few critics who asked whether our use of digital media might be damaging rather than augmenting us. Understanding these critics' contentions is an essential starting place for those of us who are trying to avoid social media damage and take advantage of digital augmentation. Foremost among these critics is Carr, whose brilliantly (if misleadingly) titled *Atlantic Monthly* article "Is Google Making Us Stupid?" and subsequent book, *The Shallows*, triggered much debate. Carr is not alone. American University linguist Naomi Baron, technology critic Maggie Jackson, former Apple and Microsoft researcher Stone, and child development specialist Maryanne Wolf have each taken different angles on the same question; each of their cautionary approaches is worthy of consideration by any social media enthusiast who claims to think critically. I agree and disagree with each of them, to different degrees and for different reasons, but I know that I've deepened my efforts to improve my social media literacies by first inquiring into the possibility that alluring new ways to know and socialize might have destructive effects. I'll take the liberty of summarizing these authors' arguments and some counterarguments, and recommend the source texts if you want to get more than the brief general descriptions I can provide here.

Digital networks, Carr believes, cause us to develop habits that ultimately harm our brains and damage culture. The literate manner of thinking that led to literature and science depended on the ability to concentrate on written works for extended periods along with the capacity for deep, contemplative, or analytic thought. Many of Carr's assertions rest on the same recent neuroscientific studies cited by those who hold that attention can be trained—the brain's ability to reconfigure itself that has become known as neural plasticity or neuroplasticity. What Carr fears is "unwelcome neuroplastic adaptations." The problem is not distractibility but instead, Carr maintains, a deep "intellectual ethic" of the Internet-using population, "a set of assumptions about how the human minds works or should work." Carr claims that an adherence to these assumptions is damaging the ability of individuals and societies to focus. It does this by inscribing the Web's innate, frenetic, and shallow sensibilities into our very neural circuitry. The Net's ethic disrupts the capacity for deep thought in a number of ways, Carr argues, starting with the elementary building block of the Web: the hyperlink. "Hyperlinks," writes Carr, "encourage us to dip in and out of a series of texts rather than devote sustained attention to any one of them. Hyperlinks are designed to grab our attention. Their value as navigational tools is inextricable from the distraction they cause." Carr fears that the hyperlink mind-set spells the death of the ways of thinking fostered by the book—and this, he feels, will be disastrous for individuals and culture.[46]

Another shallowing-out force Carr cites is Web search, which was the core of his original "Is Google Making Us Stupid?" article. "A search engine," he writes, "often draws our attention to a particular snippet of text, a few words or sentences that have strong relevance to whatever we're searching for at the moment, while providing little incentive for taking in the work as a whole."[47] Multimedia and the design of the graphical user interface in which multiple windows always present multiple opportunities for distraction are another corrosive force, in Carr's view. When the plastic brain falls into habits of indulging these media, Carr argues, the cognitive effects are profound. We lose the capacity for sustained, focused attention—"the Net seizes our attention only to scatter it." We develop "screen-based reading behavior"—nonlinear, scattered, perpetual scanning at the expense of depth and concentration. As we substitute the Web for personal memory, "we risk emptying our minds"—and as the Web makes it harder to remember, we are forced to rely on it all the more. "What we're experiencing," says Carr, "is, in a metaphorical sense, a reversal of the early trajectory of civilization: we are evolving from being cultivators of personal knowledge to being hunters and gatherers in the electronic data forest."[48]

I take issue with Carr's assumption of inevitability: a culture can choose to educate widely, as post-Gutenberg Europe and the rest of the world did, in response to a disruptive abundance of communications and ways of communicating. In academic circles, the attitude taken by Carr and other critics I consider here is called "technological determinism," and in my opinion it can be as dangerous as a lack of awareness of technology-enabled pitfalls. Humans have agency. The Web wouldn't have existed without that agency, even given the technical medium of the Internet.

I believe Carr is right to sound the alarm, however, about the potentially harmful effects of (the unmindful use of) digital, networked media. It's all happening so fast. Can cultural institutions emerge quickly to respond to technological disruptions? In particular, Carr's insight into a shortcoming of search forced me to think about how I've taught myself, my daughter, and my students to look beyond the snippet that a search query reveals. A search query, like a Wikipedia page, often is a bad place to end your inquiry, but an excellent place to start. An online knowledge search should be like cinema, not a snapshot, or a process of knowledge building, not a fast answer. Sometimes you want an answer (to, say, What year was René Descartes born?) and sometimes you want knowledge (concerning, for example, What does metacognition mean?) On reflection, I realized that I explore the context around a search by employing the very tool Carr abhors—the hyperlink. What Carr sees as a misleadingly incomplete fragment, I view as a kernel of knowledge that points to other kernels that, taken together, can reveal overarching networks, connections, and systems. Where Carr wants to avoid clicking in the first place, I've taken to clicking around to get a better sense of topics.

Engine-assisted search in itself is not a fragmenting, decontextualizing, shallowing force. Again, I reject the simple deterministic answer that the machine's affordances inevitably control the way we use the mechanism. Shallow inquiry—the uninformed way in which many people use search engines to find answers—is the deeper problem, and one that can be remedied culturally. Just as the ancient arts of rhetoric taught citizens how to construct and weigh arguments, a mindful rhetoric of digital search would concentrate attention on the process of inquiry—the kinds of questions people turn into initial search queries, and the kinds of further questions that can deepen their search. Search affords snippet thinking. Carr is right that the medium itself offers little incentive to seek depth, but it does not compel shallow, lazy thinking. I'll get into more detail about search and inquiry when I move from the literacy of attention to the art and science of crap detection—since finding what you really need to know and knowing how to sort the good from the bad info are complementary (and essential) skills in today's infosphere.

Carr's literary device of exaggeration is entertaining, but his extreme stance weakens his dismissal of the power of culture to tame media's attentional effects. To write his book, Carr did not just learn to practice judicious self-discipline. He moved "from a highly connected suburb of Boston to the mountains of Colorado," closed his Twitter account and blog, took a hiatus from Facebook, and restricted his email use. Dramatic? Undoubtedly. Necessary? I'm not so sure. He claimed his "synapses howled for their Net fix." He knows as well as I do that the world harbored abundant attractions to lure writers long before the Internet, and that the writer's primary response to the world's infinite opportunity for distraction ought to be one of internal discipline, not ascetic withdrawal. Carr found himself "sneaking clicks" or going on the occasional "daylong Web binge," until "the cravings subsided." Yes, I see what he's getting at. Who hasn't experienced the magnetic pull of social media during vacations and other times when we know it is inappropriate? But isn't the same thing true of drinking, sex, or any of a myriad of attractive pastimes? In finishing his book, he describes "backsliding" into his old habits.[49] Carr's approach keeps the reader's attention focused and brings his argument to life in a vivid way, but in so doing he ignores both potential benefits of heightened connectivity and the creative possibilities of the social media platforms he rejects.

Extremes make for good storytelling, but framing the question of how social media affects individuals as one of addiction overlooks some of its most crucial dimensions—including the all-important question, Can we learn to turn the new way of thinking into a net positive, the way humans learned to deal with the alphabetic culture that Plato warned about, through education and norms? Even Carr admits that people have power over the kinds of distraction he believes the online world offers: "Our brains are very adaptable and flexible. If you change your habits, your brain is very happy to go along. The hard thing is to change your habits."[50] I recommend paying attention to the possibilities that Carr raises, and using them as instruments for self-examination. But I don't believe Carr's abstinence-only, zero-tolerance solution can resolve these issues.

I think that the power of the Internet mind-set is up to us, just as it was in relation to the Gutenberg and alphabet mind-sets. Defending this potential, New York University professor Clay Shirky writes in "Why Abundance Is Good: A Reply to Nick Carr":

I think Carr's premises are correct: the mechanisms of media affect the nature of thought. The web presents us with unprecedented abundance. This can lead to interrupt-driven info-snacking, which robs people of the ability to find time to think about just one thing persistently. I also think that these changes are significant enough to motivate us to do something about it. I disagree, however, about what it is we should actually be doing. . . . The change we are in the middle of isn't minor and it isn't optional, but nor are its contours set in stone. We are a long way from discovering and perfecting the net's native forms, what [Roland] Barthes called the "genius" particular to a medium. To get there, we must find ways to focus amid new intellectual abundance, but this is not a new challenge. Once the printing press meant that there were more books than a person could read in a lifetime, scholars had to sharpen disciplines and publishers define genres, as a bulwark against the information overload of the 16th century. Society was better after that transition than before, even though it took two hundred years to get there. And now we're facing a similar challenge, caused again by abundance, and taking it on will again mean altering our historic models for the summa bonum of educated life. It will be hard and complicated; abundance precipitates greater social change than scarcity. But our older habits of consumption weren't virtuous, they were just a side-effect of living in an environment of impoverished access. Nostalgia for the accidental scarcity we've just emerged from is just a sideshow; the main event is trying to shape the greatest expansion of expressive capability the world has ever known.[51]

The expansion of expressive capability mentioned by Shirky is not, in itself, a wholly new phenomenon. As I'll discuss in later chapters, the human propensity as well as talent for symbolic expression has been both a biological and cultural evolutionary driver. What is happening to our language? is as important a question to ask as What is happening to our minds? and What is happening to our social relations?

In *Always On: Language in an Online and Mobile World*, Baron looks critically at the consequences of a rapidly evolving linguistic environment in which LOL (laugh out loud) and SMS (Short Message Service) seem to have created an abbreviated jargon overnight. She suggests that "email and its descendants" have triggered two fundamental changes. First, new communication technologies give us increasing control over how, when, and with whom we interact—what Baron calls "volume control." Second, as we replace much of our spoken interaction with written exchanges, Baron fears that quantity increases and quality suffers.[52] (A stronger version of Baron's claim that democratized access to publishing leads to so much crap it's killing culture is central to Andrew Keen's book *The Cult of the Amateur*.[53] I reject this argument on the grounds that educating readers how to value good writing proved to be a better solution in the age of print than the remedy attempted by monarchies: licensing publishers. The huge amount of poorly developed and badly written printed matter churned out by printing presses did not prevent Charles Dickens or Honoré de Balzac from finding widespread readerships.)

I found Baron instructive regarding specific ways social media challenge traditional definitions of sociality. Baron is right, in my opinion, to urge us all to cast a critical eye on any form of socializing that can be turned on and off at will. In my own life, volume control has been a net benefit, but it's not without its shadowy side. My craft as a writer demands that I spend my days mostly alone in a room. Given my circumstances, gaining the power to click into a virtual community increased my daily social interaction, since I was already isolated. After twenty-five years of online socializing, however, I understand (and caution others against) the danger of confining myself exclusively to communities I can click on and off. I'm healthier, and so is my society, because I'm embedded in family, neighborhood, hometown, campus, and social cyberspace. The people I've met online as well as mostly communicate with through virtual means have come to my rescue in times of peril, bought me lunch in Amsterdam and Istanbul, showed me caring, and shared the fun that any kind of community worthy of the name strives for—but I learned long ago that I also need to maintain my face-to-face connections.

Turkle sees the media affordance of Baron's volume control as a root cause of the alienating aspects of social media:

People would rather text than talk, because they can control how much time it takes. They can control where it fits in their schedule. When you have the amount of velocity and volume [of communication] that we have in our lives, we have to control our communications very dramatically. So controlling relationships becomes a major theme in digital communication. And that's what sometimes makes us feel alone together—because controlled relationships are not necessarily relationships in which you feel kinship.[54]

Understanding the ways volume control can be both a benefit and danger is a prerequisite to taming it. The social worlds I encounter through digital media are definitely real even if they aren't physical, but I've learned to resist thinking of them as my only reality. Baron concludes that change— good or bad—in language, thought, and society depends ultimately on individual choice. I share Baron's fears and cautions. My own stance toward media literacy—the reason I wrote this book—is based on the same conclusion Baron reached: that human agency, not just technology, is key. What you and I know, think, and do at this moment of technology-initiated yet human-centered change matters.

In *Distracted: The Erosion of Attention and the Coming Dark Age*, Jackson writes: "The way we live is eroding our capacity for deep, sustained, perceptive attention—the building block of intimacy, wisdom, and cultural progress. Moreover, this disintegration may come at a great cost to ourselves and to society." We should be worried, says Jackson, because "the erosion is reaching critical mass. We are on the verge of losing our capacity as a society for deep, sustained focus. In short, we are slipping toward a new dark age." She shows us her view of the history of what she calls "an attention-deficit culture," arguing that communications technologies, beginning with the telegraph, encouraged the development of a "culture of simultaneity and split-screen attention." Technology, she claims, feeds a need for "utopian ideals of connection," leading us deeper and deeper into virtual worlds and further from physical reality: "exploratory forays into unseen worlds are burgeoning into determined desire to increasingly inhabit new dimensions."[55] I would say that a society in which most people can read and write was once a utopian ideal, and that virtual worlds became important to our species when we first learned to manipulate symbols. Today's technology may be new, but using media to change (some would say expand) human consciousness at least goes back to forty-thousand-year-old cave paintings.

Jackson's book characterizes the threat posed by attention-deficit culture to various aspects of society. The habit of multitasking, amplified by technologies of distraction, is hurting our capacity for sustained attention. Unwittingly, we increasingly base our personal relationships on surveillance as opposed to trust. Books are disappearing (or will be soon); we're sacrificing our "hard-won" ability to wrestle with a text. The looming dark age Jackson worries about as the ultimate consequence of attention-deficit culture will perhaps not come in the form of a dramatic collapse of civilization but more likely is happening as a slow, sinking, unavoidable decline instead. "Mesmerized by the flickering charms and lightning-fast shifts in our own time," she writes, "perhaps we can't tell at first glance whether what's creeping around us are rippling shadows or a fearful twilight."[56] I see the same possible chasm that Jackson foretells. I'm looking for ways to climb out. I don't see myself arguing with Jackson; rather, I see myself attempting to answer her challenge.

Jackson and I see eye to eye that such a catastrophe potentially might be forestalled by paying attention to attention. Despite her semiapocalyptic predictions, Jackson ends optimistically: "And yet, a *renaissance of attention* may be at hand. An antidote to our epidemic distraction lies in a set of astonishing discoveries: attention can be understood, strengthened, and taught. However we may define progress now or in the future, there is a spark of hopefulness in that."[57] When I ask myself what is to be done if Jackson is even partially right, I conclude that teaching people how to practice more mindful mediated communication seems the most feasible remedy. I like Jackson's query in an excellent Boston.com article about attention training: "If focus skills can be groomed, as research has begun to hint, the important next question is whether, and how, attention should be integrated into education. Will attention become a 21st-century 'discipline,' a skill taught by parents, educators, even employers? Already a growing number of educators are showing interest in attention training, mostly through the practice of meditation in the classroom."[58] I'm with Jackson; self-control along with the skillful use of attention, participation, crap detection, collaboration, and network awareness through social media ought to be taught to future netizens as early as possible.

Stone, in addition to identifying email apnea, came up with another useful concept for a behavior that digital media enables far more effectively than olden media did: communicating with multiple people and seeking information from multiple sources simultaneously. Who doesn't recognize the detectable hesitation in the voice of the person you are talking to on a telephone when you just know they are surfing the Web or checking their email? This antisocial form of multitasking seems to pop up everywhere, from laptops in classrooms to BlackBerries in meetings. The *New York Times* quoted Eric Schmidt, CEO of Google, in an article: "Shortly after joining the company and its young founders, Sergey Brin and Larry Page, he [Schmidt] was frustrated that people were answering e-mail on their laptops at meetings while he was speaking. 'I've given up' trying to change such behavior, he says. 'They have to answer their e-mail. Velocity matters.'"[59]

Stone names this behavior "continuous partial attention, an always-on, anywhere, anytime, any place behavior that involves an artificial sense of constant crisis." She sees the smart-phone-and-laptop-using world seized by a new "dominant attention paradigm," characterized by what she calls "semi-sync" communication, somewhere between synchronous and asynchronous, with varying degrees of simultaneous or overlapping connectivity for different social circumstances—phone calls for close friends, text messaging for groups of friends and casual friend-to-friend conversations, and social networks for the wider orbit.[60]

As with email apnea, Stone's approach is prescriptive. Continuous partial attention can hamper opportunities for reflection and authentic social connection as well as threaten personal health and well-being. Stone's solution is twofold. First, *breathe*; email apnea appears to be a symptom of being in a state of continuous partial attention, and the treatment is the same. Second, we need to learn to manage our attention and mitigate the impulse to constantly connect. Carefully managed attention is more engaged than frantic attention, Stone suggests. She gets no argument from me! When I asked her about the specifics of learning to exercise better executive control over attention, Stone reminded me that "intention is the fuel for attention." Intention and setting goals are different, she told me, because "a goal is outside and in the future, but an intention is inside you and very present. And when does behavior change? It changes in the present." Stone isn't on a crusade against multitasking or continuous partial attention. "The most wondrous mind to me," she explained, "is the resilient, flexible mind that has a capacity to adopt any kind of attention strategy and a sensibility to determine which kind of attention matches the present situation."[61]

When I looked for scientific means of sorting through the theorizing and philosophizing about whether the Internet is making us stupid or not, I was thrilled to discover Wolf, a professor of child development at Tufts University. An expert on the neural, cognitive, and cultural components of reading, she confronts the issue of how media such as books or computer networks affect the brain. As Wolf observes in her book *Proust and the Squid*:

We were never born to read. Human beings invented reading only a few thousand years ago. And with this invention, we rearranged the very organization of our brain, which in turn expanded the ways we were able to think, which altered the intellectual evolution of our species. Reading is one of the single most remarkable inventions in history; the ability to record history is one of its consequences. Our ancestors' invention could come about only because of the human brain's extraordinary ability to make new connections among its existing structures, a process made possible by the brain's ability to be shaped by experience. This plasticity at the heart of the brain's design forms the basis for much of who we are, and we might become.[62]

Human thinking processes are neither wired nor rewired, although it is convenient to think of them in that way. Even if the probability that a specific set of brain cells will fire in synchronization does resemble fixed circuitry, the brain works in a more dynamic way than the wiring metaphor implies. Wolf emphasizes that groups of neurons create new connections and strengthen pathways between them in specific networks whenever a person acquires a new skill. "Thanks to this design," Wolf notes, "we come into the world programmed with the capacity to change what is given to us by nature, so that we can go beyond it."[63] This is the neural plasticity feared by Carr.

Humans are born biologically equipped to recognize visual patterns, then extract meaning from them, but meshing language, vision, and attention to transmit knowledge was invented; reading is a mind technology. The visual component of reading appears to make use of deeply embedded perceptual mechanisms that probably evolved in order to track predators and prey by deciphering their footprints.[64] Each of the innate capacities recruited by the brain for use in the reading process evolved in response to some survival requirement. Wolf puts it this way:

It would seem more than likely that the reading brain exploited older neuronal pathways originally designed not only for vision but for connecting vision to conceptual and linguistic functions: for example, connecting the quick recognition of a shape with a rapid inference that this footprint can signal danger; connecting a recognized tool, predator, or enemy with the retrieval of a word. When confronted, therefore, with the task of inventing functions like literacy and numeracy, our brain had at its disposal three ingenious design principles: the capacity to make new connections among older structures; the capacity to form areas of exquisitely precise specialization for recognizing patterns of information; and the ability to learn to recruit and connect information from these areas automatically. In one way or another these three principles of brain organization [are] a foundation for all of readings evolution, developments, and failure.[65]

Reading requires these separate perceptual and cognitive functions to work in a highly coordinated manner. The visual recognition of a letterform, the sequencing of letterforms, and recognition of sequences as representations of words all must synchronize in ways that our brains did not acquire through biological evolution. As Harvard psychologist Steven Pinker describes it, "Children are wired for sound, but print is an optional accessory that must be painstakingly bolted on."[66] By training humans to read, we can harness these naturally uncoordinated processes to create a valuable new metaprocess—the ability to encode knowledge in written marks, transmit those marks across time and space, and for any other trained person a thousand miles away or a thousand years in the future to decode that knowledge. The display of letterforms on a sign and the rote recitation of the alphabet by students—tools that go back to the dawn of civilization—are brain-changing methods for creating entirely new mental mechanisms by connecting existing ones in a novel way.

Another reading expert, Dehaene, calls this adoption of previously evolved brain mechanisms to serve new tasks "neuronal recycling."[67] Multiple levels of neural and cultural plasticity make the progressive bootstrapping power of literacy possible: training humans to read induces permanent changes in the way brain networks function; preexisting brain functions can be coordinated to perform the new tasks such as reading, writing, and arithmetic; and literacy-augmented populations can invent better alphabets, printing presses, and wireless telegraphs. Understanding the broader dynamics surrounding the encounter of your brain and the Web is essential metaknowledge that will lend power to your efforts to gain greater control over mediated attention.

Wolf addressed the concerns raised by Socrates twenty-five hundred years ago in regard to the alphabet. Wolf shares Socrates's objections, which have the advantage over Wolf's conjecture of being interpretable in light of the history of literacy:

Examining written language, Socrates took a stand that usually comes as a surprise: he felt passionately that the written word pose[s] serious risks to society. . . . And as we examine our own intellectual transition to new modes of acquiring information, these objections deserve our every effort to get to their essence. First, Socrates posited that oral and written words play very different roles in an individual's intellectual life; second, he regarded the new—and much less stringent—requirements that written language placed both on memory and on the internalization of knowledge as catastrophic; and third, he passionately advocated the unique role that oral language plays in the development of morality and virtue in the society. In each instance Socrates judged written words inferior to spoken words, for reasons that remain powerfully cautionary to this day. . . . Ultimately, Socrates did not fear reading. He feared superfluity of knowledge and its corollary—superficial understanding. Reading by the untutored represented an irreversible, invisible loss of control over knowledge. As Socrates put it, "once a thing is put in writing, its composition, whatever it may be, drifts all over the place, getting into hands not only of those who understand it, but equally of those who have no business with it; it doesn't know how to address the right people, and not address the wrong. And when it is ill treated and unfairly abused it always needs its parents to come to its help, being unable to defend or help itself." Underneath his ever present humor and seasoned irony lies a profound fear that literacy without the guidance of the teacher or of a society permits dangerous access to knowledge. Reading presented Socrates with a new version of Pandora's box: once written language was released there could be no accounting for what would be written, who would read it, or how readers might interpret it.[68]

Powerful technologies always entail trade-offs and while the power of a new tool is evident early, the prices we pay may take longer to become visible. Socrates was concerned about knowledge in the hands of the untutored. Many slaves supported each free Athenian of Socrates's time, and slaves were barred from learning the "liberal arts"—the communication skills necessary for the citizens of a democracy to retain their liberty. Mass literacy and education, enabled first by the alphabet and then by the printing press, made possible the forms of knowledge and governance we know today. One price of mass literacy is that not everyone who learns to read has personal access to Socrates. If you want more than the elite to be free and autonomous, you face the problem that there aren't enough tutors to explain the meaning of everything. If you want to radically broaden the scope of literacy, the loss of some depth might be part of the price. The question at present, not a new one, is whether loss of depth is preventable or instead is an inevitable consequence of technologies that democratize knowledge.

Democratization enables vulgarization. As cultural practices become more common, they also become more coarse and misinterpreted. In the early twentieth century, the young print journalist Walter Lippmann claimed that U.S. citizens are too gullible and ill informed to govern a modern, complex society. In response, philosopher-activist John Dewey responded that in a democracy, the answer was not, as Lippmann suggested, to confine governance to an elite but rather to make the entire population less gullible through better public education and better informed through better journalism.[69]

I hear echoes of Lippmann in Carr's arguments and see elements of Dewey's response in Wolf's proposals. Wolf considers the persistence in human populations of dyslexia, the perceptual anomalies that make it difficult for some people to learn to read, to be "a daily evolutionary reminder that very different organizations of the brain are possible. Some organizations may not work well for reading, yet are critical for the creation of buildings and art and the recognition of patterns—whether on ancient battlefield or in biopsy slides. Some of these variations of the brain's organization may lend themselves to the requirements of modes of communication just on the horizon." At the same time, Wolf is cautious about the cognitive effects of the increasing speed with which information is created, processed, and consumed. Intellectual skills flourished, she writes, because of "the secret gift of time to think that lies at the core of the reading brain's design."[70]

Having helped train her dyslexic son to read, plus having studied dyslexia scientifically, Wolf appears to be a strong believer in the power of teaching and learning. She contends that the demonstrable power of teaching alphabetic literacy can be applied to the challenge of information and media literacies:

We must teach our children to be "bitextual" or "multitextual," able to read and analyze texts flexibly in different ways, with more deliberate instruction at every stage of development on the inferential, demanding aspects of any text. . . . My major conclusion from an examination of the developing reader is a cautionary one. I fear that many of our children are in danger of becoming just what Socrates warned us against—a society of decoders of information, whose false sense of knowing distracts them from a deeper development of their intellectual potential. It does not need to be so, if we teach them well, a charge that is equally applicable to our children with dyslexia.[71]

Developing a pedagogy of attention is, I believe, the basis for Wolf's kind of education.

Mindfulness in an Always-On World

Mindfulness in its most general sense is about waking up from a life on automatic, and being sensitive to novelty in our everyday experiences. With mindful awareness the flow of energy and information that is our mind enters our conscious attention and we can both appreciate its contents and also come to regulate its flow in a new way. Mindful awareness, as we will see, actually involves more than just simply being aware: it involves being aware of aspects of the mind itself. Instead of being on automatic and mindless, mindfulness helps us awaken, and by reflecting upon the mind we are enabled to make choices and change becomes possible.

How we focus attention helps directly shape the mind when we develop a certain form of attention to our here-and-now experience and to the nature of our mind itself, we create a special form of awareness, mindfulness.

—Daniel J. Siegel, *The Mindful Brain*, 2007

Scientists and meditators both claim that simple exercises can increase attentional agility. That's why the Chicago Bulls and Los Angeles Lakers use a meditation teacher. Attention processes, like muscles, can be strengthened through exercise, resulting in measurable changes in brain functions.[72] Here's how it works, and how to do it.

I've been interested in the neuroscience of attention training since 1968, when I based my undergraduate thesis at Reed College on the work of Joe Kamiya, PhD, University of California at San Francisco.[73] Kamiya had studied the electroencephalographic (more popularly known as brain wave) patterns of Buddhist monks and claimed that these experienced meditators exhibited a higher than normal incidence of the eight to ten cycles per second alpha wave. The electric rhythms detectable by scalp electrodes can't resolve a fine-grained image of brain activities; electroencephalogram (EEG) study is like analyzing the seabed by standing on a cliff and looking at the waves break on shore. Yet brain waves do offer a window on the neural correlates of states of mind. In 1968, an exciting idea began to energize research activity: making people conscious of their brain waves can enable them to influence their brain wave output. Kamiya demonstrated that ordinary subjects who were not meditation practitioners could learn to increase their proportion of alpha wave activity by playing an audible tone to subjects whenever alpha waves were detected and instructing subjects to maintain the sound as much as possible—a procedure that came to be known as "brain biofeedback" or "neurofeedback."[74]

After graduating from Reed, I studied at the State University of New York at Stonybrook with Professor Lester Fehmi, whose laboratory was exploring the possibilities of Kamiya's findings. Fehmi continues his research to this day, but I left neuroscience in 1970. Fehmi's most recent book, *Open Focus: Harnessing the Power of Attention to Heal Mind and Body*, published in 2007 with coauthor Jim Robbins, lends a descriptive name to the mental state associated with brain wave control.[75] The state of mind associated with alpha, as I recall it, was indeed focused in that it felt the opposite of dreamy, wandering, or distracted. But it was, well, open because it wasn't focused on any one thing—including trying to make the sound of the tone by emitting more alpha waves. I spent many hours in an electromagnetically shielded room at the university, where I affixed electrodes to the scalp of our subjects (and myself) and converted our EEG signals into wavy lines inked onto long, fanfolded strips of paper. It was very analogue; there was nothing digital about it. I learned to recognize the alpha pattern by looking at the paper as it scrolled out of the machine.

I recall that I was best able sustain the audible feedback tone and inscribe the right kind of squiggles onto the paper by *not* making an effort. The feeling is more of a "letting go." I would stop making a mental effort, and then the tone and squiggles would flow. Fehmi's open focus is a good descriptor of a kind of attention that is not directing itself at a single task, or maintaining a narrative linking multiple perceptions, thoughts, plans, goals, and memories, but that is continuously awake and alert to itself in the present. After I left the State University of New York, I tried to raise money for an electronics wizard friend of mine to create a brain biofeedback unit in a briefcase. I remember thinking that my friend was joking when he told me that he was going to abandon months of circuit design. "Here is the amplifier circuitry I was trying to get into a briefcase," he said, holding up a little metal gizmo about the size of a pencil eraser. It was an "operational amplifier," one of the first successful uses of the newly invented integrated circuitry that miniaturized complex circuits on silicon chips.

As Fehmi and other neurofeedback practitioners continue to prove, learning to control brain waves and other bodily signals can indeed be useful in learning meditation, managing pain, and even controlling computers (you can be certain whenever you read a headline about scientists "controlling computers with thoughts" that it concerns neurofeedback). Despite the persistent interest of a few cognitive scientists and clinicians along with a somewhat New Agey cult of enthusiasts, neurofeedback has not made the multimillionfold progress over the past four decades that miniaturized digital circuitry has.[76] Miniaturized digital circuitry, however, also made possible the more sophisticated brain-imaging devices such as functional magnetic resonance imaging (fMRI) that are being used to observe finergrain brain correlates of mental processes. Although the brain mechanisms underlying attentional processes are complex, subsequent research has strongly reinforced Kamiya's finding that becoming aware of attention processes through their neural correlates can help people learn to control those processes. What excited me in 1968 still excites attention researchers today—evidence that changing the mind can lead to changing the brain.

Both neuroscientific and contemplative schools of thought about attention training use the same word to describe the mental self-leverage that makes it possible for the mind to change the brain: mindfulness. In the words of mindfulness teacher Jon Kabat-Zinn, mindfulness is "the awareness that emerges through paying attention on purpose."[77] That awareness, which even tentative direct experimentation can grant to some noticeable degree, is the power tool that all the other literacies depend on. Mindfulness is what connects your attention to skills of digital participation, collaboration, crap detection, and network smarts. Deliberately exercised, continually strengthened, and judiciously applied, mindfulness is the most important practice for anyone who is trying to swim through the infostream instead of being swept away by it.

Soren Gordhamer, author of *Wisdom 2.0: Ancient Secrets for the Creative and Constantly Connected*, wrote of mindfulness:

A teacher of mine used to say, "Do what you normally do; just do it with aware-ness." He meant that if you bring consciousness to any activity, your experience of it changes. If you make conscious a conversation, you better know what to say and can more fully listen to the other person; if you make conscious your work on a project, you can more easily see what it needs to progress; if you make conscious an unskillful habit, you can better understand why you follow it and how to release it. If you add consciousness to any activity, the nature of it changes, and the creative is illuminated.[78]

Gordhamer interviewed George Mumford, the sports psychologist who taught mindfulness meditation to the Bulls and Lakers. I found Mumford's reply instructive:

In sports, what gets people's attention is this idea of being in the zone, or playing in the zone. When they are playing their best, they can do no wrong, and no matter what happens they are always a step quicker, a step ahead. That happens when we are in the moment, when we are mindful of what is going on. There's a lack of self-consciousness, there's a relaxed concentration, and there's this sense of effortless-ness, of being in the flow. We have that experience in other parts of our life, but we equate it with sports because there are rules and guidelines, and it is a situation where you get immediate feedback. When we are in the moment and absorbed with the activity, we play our best. That happens once and awhile, but it happens more often if we learn how to be more mindful. By mindful, I mean being aware, being engaged with the present moment. Mindfulness is useful because it is through this that we can see what is going on. It means knowing what needs to happen and doing it.[79]

When using the mind to influence the brain, it helps to start by know-ing why you should believe these practices work. A number of studies—more than I can cite here succinctly—have demonstrated that mindfulness exercises can have lasting effects on attention. After a brief look at the experimental evidence, I'll get down to practical basics: that is, what other researchers and I have found to be successful in our efforts to become more mindful digital citizens. Fortunately, you don't have to take the advice that follows on faith alone—even faith in experimental data. Your best proof ultimately is to try it yourself. You've already started influencing your attention by thinking about it, or as noted above, by what some learning theorists call metacognition.

The word metacognition means the act of thinking about thinking and more; "metacognitive strategies" enable people to apply what they have learned about attention control to new learning tasks—a "higher order thinking which involves active control over the cognitive processes engaged in learning."[80] Wikipedia has an excellent page on metacognition, which notes that "metacognition is classified into three components":

1. *Metacognitive knowledge* (also called metacognitive awareness) is what individuals know about themselves and others as cognitive processors.
2. *Metacognitive regulation* is the regulation of cognition and learning experiences through a set of activities that help people control their learning.
3. *Metacognitive experiences* are those experiences that have something to do with the current, ongoing cognitive endeavor.[81]

I'm introducing metacognition at this point because education theorists claim that learning about metacognition can lead to doing it more effectively—and emphasize that learning about and doing it yourself are both necessary. Jennifer A. Livingston, professor of educational psychology at the State University of New York at Buffalo, writes:

While there are several approaches to metacognitive instruction, the most effective involve providing the learner with both knowledge of cognitive processes and strategies (to be used as metacognitive knowledge), and experience or practice in using both cognitive and metacognitive strategies and evaluating the outcomes of their efforts (develops metacognitive regulation). Simply providing knowledge without experience or vice versa does not seem to be sufficient for the development of metacognitive control.[82]

"Knowledge" is an abstraction; whatever it is, it's rooted in the activities of the brain. Neuroscientists are using new tools to look at the neural substrates of attention, thought, and knowledge. Meditation practice turns out to be a highly desirable experimental variable for those trying to study whether attention training can improve attentional agility.

Norman Farb and six other scientists at the University of Toronto used fMRI visualization to study novices and subjects who had practiced mindfulness meditation techniques. (I'll get into the specifics of mindfulness meditation shortly, but I can describe it here simply enough: with a nonjudging attitude, pay attention to your breath and return your attention to it when you find your mind wandering; repeat this process over and over again.) The Toronto neuroscientists claim to have discovered that people use two different kinds of attention that correspond to different brain networks. One network, which the researchers called "the default network" and more evocatively "the narrative network," is what is active when you think about yourself and your fantasies, evoke memories, and formulate plans.[83]

David Rock, in a *Psychology Today* article titled "The Neuroscience of Mindfulness," explains:

When the default network is active, you are thinking about your history and future and all the people you know, including yourself, and how this giant tapestry of information weaves together. . . . [W]hen you experience the world using this narrative network, you take in information from the outside world, process it through a filter of what everything means, and add your interpretations. Sitting on the dock with your narrative circuit active, a cool breeze isn't a cool breeze, it's a sign that summer will be over soon, which starts you thinking about where to go skiing, and whether your ski suit needs a dry clean. The default network is active for most of your waking moments and doesn't take much effort to operate. There's nothing wrong with this network, the point here is you don't want to limit yourself to only experiencing the world through this network.[84]

The other brain network and accompanying attentional state identified by Farb and his colleagues is a "direct experience network," during which an entirely different set of brain regions are active. The narrative network is essential for planning, and plans can create perceptual frames that allow you to focus and also filter out irrelevant information. But direct experience enables more sensory information to be perceived in the present and thus a more flexible response (that is, the opposite of a habitual or compulsive response). The narrative network is what is most active when your mind drifts into plans and memories during meditation. The direct experience network is what is most active when you keep your attention on your breath; it is probably the one that is active when an athlete is in the zone.

A multitude of research teams have tested the effect of mindfulness meditation on the kind of attention networks demonstrated by Farb and his colleagues. Amisha Jha, Jason Krompinger, and Michael Baime in the Department of Psychology at the University of Pennsylvania examined neural and cognitive attentional subsystems, both before and after mindfulness training, concluding that "mindfulness training may improve attention-related behavioral responses by enhancing functioning of specific subcomponents of attention."[85] The *New York Times* reported that Jha, Krompinger, and Baime have been engaged in training U.S. Marines. With streams of real-time, life-and-death information from drone aircraft and other sources flooding display screens, attention training is a serious military issue: "At an Army base on Oahu, Hawaii, researchers are training soldiers' brains with a program called 'mindfulness-based mind fitness training.' It asks soldiers to concentrate on a part of their body, the feeling of a foot on the floor or of sitting on a chair, and then move to another focus, like listening to the hum of the air-conditioner or passing cars."[86]

Psychologist Michael I. Posner and science writer Brenda Patoine observed, "Given the importance of the executive attention network, my colleagues and I wondered what might improve its efficiency. To find out, we adapted a series of exercises, originally designed to train monkeys for space travel, to investigate the effects of attention-training exercises in 4-6 year old children," and discovered that "tasks specifically designed to exercise the underlying networks can indeed improve attention, and that this kind of training can translate to better general cognition."[87] Pamela D. Hall, assistant professor of psychology at Barry University, claims that college students who learned meditation had significantly higher grade point averages compared to the control group.[88]

There is more research into the neuroscience of meditation, mindfulness, and attention training than I have room to summarize here. Branch out from the bibliographies of the work I've cited and you can explore the scope of this burgeoning young field for yourself. One more study is worth noting. In the late 1960s, Stanford University psychology professor Walter Mischel conducted a simple experiment with children as young as four. He introduced the subjects to a marshmallow, which he put on a table in front of them, informing the young subjects that they could eat the marshmallow, but if they could wait while the experimenter stepped out of the room for a few minutes, they could have two marshmallows. Mischel filmed the subjects when they were alone with the temptation. Some of them ate the marshmallow, forfeiting the second. Some of them waited for the extra reward. The research became more interesting years later, when Mischel tracked down his former subjects. He discovered that those children who couldn't wait to eat the marshmallow were more likely to have behavioral problems and ended up with Scholastic Assessment Test scores that were on average 210 points lower than those who had experienced patience decades before. Mischel concluded that "intelligence is largely at the mercy of self-control," and more specifically, the self-control of *attention*.[89]

Jonah Lehrer, in his *New Yorker* article about Mischel's research, writes:

Mischel's conclusion, based on hundreds of hours of observation, was that the crucial skill was the "strategic allocation of attention." Instead of getting obsessed with the marshmallow—the "hot stimulus"—the patient children distracted themselves by covering their eyes, pretending to play hide-and-seek underneath the desk, or singing songs from "Sesame Street." Their desire wasn't defeated—it was merely forgotten. "If you're thinking about the marshmallow and how delicious it is, then you're going to eat it," Mischel says. "The key is to avoid thinking about it in the first place." In adults, this skill is often referred to as metacognition, or thinking about thinking, and it's what allows people to outsmart their shortcomings. (When Odysseus had himself tied to the ship's mast, he was using some of the skills of metacognition: knowing he wouldn't be able to resist the Sirens' song, he made it impossible to give in.)[90]

Mischel believes that self-control is more than an innate ability, but also incorporates environmental and individual variables such as motivation, perspectives, and heuristics (mental tricks) employed to maximize rewards

and minimize complications (which are states, not traits). "We can't control the world," remarks Mischel, "but we can control how we think about it."[91] If a four-year-old can achieve the miraculous feat of delayed gratification with a few simple tips, there may be enormous payoffs to becoming students of our own minds. Undoubtedly, there are many tricks just as easy as "look away" that may allow us to maximize our attentional potential.

Now let's look at how you can test attention training for yourself. I'll start by demystifying meditation—an exercise that many who haven't tried might imagine to be far more difficult than it really is (at least at the beginning). In addition to practical exercises like mindfulness meditation, I'll introduce some strategic goal-achieving tips that I've gathered over the years from others who are more adept.

Training the Puppy Mind

Breath is the bridge which connects life to consciousness, which unites your body to your thoughts. Whenever your mind becomes scattered, use your breath as the means to take hold of your mind again.
—Thich Nhat Hanh, *The Miracle of Mindfulness*, 1975

Breathe in. Breathe out. (My ankle is going to start bothering me in this position [body thought].) Breathe in. (How will I describe this? [planning]) Breathe out. Breathe in. (I wasn't so fast to jump from thought to thought when I did this yesterday [remembering].) Breathe in. Breathe out. Breathe in.

Continue in this manner for a few pages and you get the picture of what goes on in the inner theater of my awareness when I meditate. I itch, fret, free-associate, and think about thinking about thinking. I sit with my back straight, legs crossed, eyes half closed, hands held in front of my navel, one hand resting in the other palm, and focus my attention on my breath, observing my body pull air in, then expel it, and watching my mind as it attends to my breathing and drifts involuntarily to one thought or another. Whenever I become aware that my intention to concentrate on my breath has been distracted by the emergence of a thought or feeling, I label the thought ("body thought," "planning," "remembering," etc.) and then let it drift away on its own, just the way it emerged. I bring my attention back to my breath. I don't try to "think about nothing." I don't strive to do better than I did yesterday or last year. I simply observe the way thoughts emerge and pass away with or without my conscious intent.

If you haven't done it, watching your breath with your eyes closed and labeling your thoughts as they pass through your mind sounds like a colossal waste of time. I admit that I get antsy, and look forward to getting back to work, play, or whatever I had been doing. I don't assign the "fun" tag to meditation. But this simple exercise provided an invaluable foundation when I started striving to surf the Web without becoming distracted (or more precisely, without becoming *too* distracted—for me, distraction online often leads to discovery, and only becomes a problem when it grows compulsive and saps time from a more important task). I've learned to use exercises in focusing awareness of my own breathing—as simply as I've described it—as a tool for learning to control my attention. Every day before lunch, I meditate for ten or fifteen minutes. I've been doing this for more than ten years now. Even though I don't meditate for long stretches, the practice of watching my mind for a small part of each day has had an observable effect. The exercise, an introspective equivalent to other rote learning exercises like reciting the alphabet or memorizing multiplication tables, creates a kind of internal observing faculty that I did not have before—a faculty that sometimes wakes up and warns me when I get caught in obsessive thought loops.

I remember distinctly in gut emotion, if not sensory detail, the first time I noticed my self-created internal observer. I was taking a kayaking class that involved two days in still water for basic training before going downstream as an expedition. I was just too top heavy and lacked sufficient upper-body strength to do the "Eskimo roll" that every kayaker has to learn in order to right the canoe when it capsizes—a tricky combination of moves that one is obliged to execute upside down in a flowing river. When the instructor told me sorrowfully that "big guys" like me often have a problem and that I'd have to paddle one of the inflatable kayaks, I took it hard. Two minutes later, I noticed that I was replaying the big-guy line over and over again to myself. I was getting into a thought loop, my internal observer noted—one that triggered unpleasant emotions and interfered with enjoying my stupendously beautiful surroundings. I was going to have to go down the river in any case, and nobody but me really cared that I was paddling a kiddie kayak, unless I wanted to devote myself to caring. I might as well savor the rest of the river trip. When I got home, I went on a campaign to lose thirty pounds.

I haven't tried to learn kayaking again, but that experience of catching myself in an obsessive and emotionally unpleasant thought was a powerful lesson to me in the value of cultivating an internal observer—a lesson that comes in handy moment to moment when I'm online.

Learning (and cultivating an internal observer is definitely a kind of learning) changes neuronal structure by strengthening connections between networks of neurons, a concept originally postulated by Freud and later reformulated by Canadian behavioral psychologist Donald O. Hebb. In 1949 Hebb proposed that learning linked neurons in new ways. He proposed that when two neurons fire at the same time repeatedly (or when one fires, causing another to fire), chemical changes occur in both, so that the two tend to connect more strongly. Hebb's concept—actually proposed by Freud sixty years before—was neatly summarized by neuroscientist Carla Shatz: "*Neurons that fire together wire together.*"[92]Hebb also demonstrated that certain kinds of associative learning led to the strengthening of coordinated firing among specific networks of neurons. This kind of neuroplastic learning by which the repeated practice of a mental skill leads to specific changes in brain functions is offered by advocates as recent scientific evidence for what meditative disciplines have asserted for millennia.[93]

When I intentionally aim the beam of my self-awareness on my breath, I strengthen networks of brain cells that eventually begin to function together even when I'm not consciously meditating. Being aware of my own thoughts, I realized, is a skill that can be learned through patient repetition. Some meditation teachers refer to the process as resembling that of training a puppy to not poop on the rug. Patiently, whenever the puppy deposits something on the rug, you pick it up and place it on the piece of newspaper near the door. Then you do it again and again. The puppy eventually starts pooping on the paper. Then you can put the paper outside. I swear that this is true, at least through the beginning levels of meditation training: much of meditation training is exactly like training a puppy, just as the teacher told me. Once I repeated the exercise for a sufficient number of days in a row, the mental pattern grooved in enough to awaken my internal observer even when I wasn't sitting in meditation.

Siegel believes that awareness of the breath performs a kind of mental-neural synchronization:

Breath is a fundamental part of life. Breathing is initiated by deep brainstem structures and is impacted directly by our emotional states. Yet breath can also be intentional. And for all of these reasons, breath awareness brings us to the heart of our lives. We come to the borderline between automatic and effortful, between mind and body. Perhaps for each of these reasons, pathways toward health included the mindful focus on the breath as a starting point on the journey.[94]

Twenty years ago, I coauthored a book with a scientist who studied sleep and dreaming—specifically the kind known as "lucid dreaming" in which you realize that you are in a dream state and begin to direct it.[95] While writing the book, I needed to train myself to recognize that I was dreaming and then seize control of the plot. The method my coauthor, Stephen LaBerge, suggested worked well: I wrote the words "Am I dreaming?" on a small card and put the card in my pocket. Throughout the day, at least once during every waking hour, I pulled the card out, looked at it, and thought about the question. About three weeks into the exercise, I pulled out the card on an otherwise perfectly ordinary day and noticed that the letters seemed to scramble to get back into place when I looked at the card, as if the letters had been loafing. I had trained myself to question whether I was awake or not at least once an hour, and without consciously intending to do it, I started reality testing while I was dreaming. Although everything and everyone around me looked real, I decided to test my perceived reality by pushing off the ground with my toes. When I began to float above the ground, I realized I was dreaming. Lucid dreaming training, like training a puppy, is not that different from the kind of intention-attention awareness I encourage you to try when you are online. Instead of "Am I dreaming?" try "Have I drifted?"

University of Washington professor David Levy has been studying the interface of attention and information for a long time. Previously, at the fabled PARC, Levy believes he may have witnessed the first email interruption, when a researcher who was demonstrating the first graphical user interface answered an email in the middle of it.[96] Levy created a college course called "Information and Contemplation."[97] In addition to teaching breath awareness to his students, Levy asked them to keep a log of their email behavior for a week, noting how their body and emotions felt, and how they were breathing while they were online. After a week of logging their interior observations during online activities, he asked the students to "look for regularities" in their logs. "Every single student discovered," Levy said in a talk to Google employees, "by doing this form of mindfulness practice, that there were certain things happening for them around email that were actually not what they wanted at all."[98]

Mindfulness in all its forms and applications certainly is an end in itself, but practicing mindfulness in regard to online attention serves a specific strategic goal. Your goal and mine in this context is not just the control but also the management of attention. Stone captures it well in writing about continuous partial attention:

We have focused on managing our time. Our opportunity is to focus on how we manage our attention. We are evolving beyond an always-on lifestyle. As we make choices to turn the technology OFF, to give full attention to others in interactions, to block out interruption-free time, and to use the full range of communication tools more appropriately, we will re-orient our trek toward a path of more engaged attention, more fulfilling relationships, and opportunities for the type of reflection that fuels innovation.[99]

In the *Huffington Post*, Stone offers valuable advice on "how to switch from managing time to managing attention":

1. Each evening or morning before you start your day, make a short list of your intentions (the result and feeling of something you want) for the day and by each, write the related to do's for that day. Try to keep your list to 5 intentions. Consciously choose what you will do and what you will not do. Keep a different list of what you will review for inclusion on other days.
2. List only what you really expect to do that day. As other things come to mind, write them on a separate list. By putting these items on a separate list, you are creating the space to be in the moment with each of your day's priorities. Review that list as you plan for the next day and determine how they fit in to your plans. Give yourself some down time, enjoy your successes at the end of the day.
3. Give yourself meaningful blocks of uninterrupted time to focus on each intention. Turn OFF technology each day during those blocks and focus on your intentions.
4. At home, be clear about what technology you'll use and where. Computer in the kitchen? Maybe not.[100]

I have information windows on my thirty-inch computer screen, but two paper index cards lie on my desk, at the periphery of my attention as I write this. On one card is my hand-sketched calendar that shows how long I need to take on finishing each chapter of this book in order to reach the deadline. Next to it is the goal I scribble out with an analog pen on the other index card. Today, for example, my index card said: "Finish 'Training the Puppy Mind.'" I don't do anything except take twenty seconds to write out the day's goal. But my purposes are served by both the object in the periphery of my attention and the metacognitive knowledge that I placed a piece of paper at that edge to help me deploy my attention more mindfully online.

Writing a goal can, and in this exercise should, become the first step in setting your intention. Strategic adviser Peter Bregman recommends an "18-Minute Plan for Managing Your Day" that takes goal-managed attention to the next level of commitment. After setting your plan for the day, Bregman suggests setting an alarm to remind you to refocus, take a deep breath, and evaluate your activity in light of your goal. Review your day at the end of it. "The power of rituals is their predictability," Bregman reminds us.[101]

The two complementary components of the goal-setting rituals recommended by Stone, Bregman, and others consists of two parts: the goal setting itself, and the deliberate creation of a ritual of setting goals. Professor B. J. Fogg, director of Stanford's Persuasive Technology Lab, puts it this way: "Forget about 'decisions'—habits are about strategy (having a smart plan) and rehearsal (until behavior is simple and automatic). Three steps to new habits: Make it tiny, find a spot, train the cycle."[102] What could be tinier than writing a one-sentence goal for the day before you get online, and reviewing it once or twice? Doing it before plunging into social media is what Fogg means by "find a spot," and "train the cycle" simply refers to doing it over and over again. In addition to practicing basic attention-to-breath meditation off-line, my recommendation for learning mindful use of social media is to establish a new habit that connects—however thinly at first—your goals to your moment-by-moment stream of attention. And remind yourself to breathe consciously from time to time.

When I interviewed Nass, he proposed that a better way for getting things done than multitasking all day is to deliberately work on a single task for fifteen to thirty minutes before going with the multitasking flow for five to ten minutes. This insight is the basis for a simple attention-training methodology known as the "Pomodoro Technique."[103] The method is easy enough. Write down your major tasks to accomplish each day on a piece of paper. Set the timer (which resembles a tomato; hence pomodoro) for twenty-five minutes and work on one task in whatever medium the task requires until you hear the alarm sound. Then take five minutes to do what you want. Repeat. Every four pomodoros, take a longer break. Train yourself to be present and aware of whether what you are doing online is going to help you achieve your own goal. Eventually you don't need the alarm clock.

Once you've evoked your attention's attention and you've started regrooving your attentional habits, you need to turn your attention to the content of your attention: How do you know that the information you've found is true, false, plausible, or preposterous? The next task for your attention in training is to develop the most important skill mindful digital citizens require to make it through life successfully these days: crap detection.

2 Crap Detection 101: How to Find What You Need to Know, and How to Decide If It's True

Every man should have a built-in automatic crap detector operating inside him.
—Ernest Hemingway, 1965

Like attention, you learn crap detection by trying out a few techniques such as those I suggest in this chapter—and then putting them into practice. Then make them habitual. The danger of distraction or credulity is made possible by digital media, but the danger you and I can do something about is the bad habit of not controlling attention or failing to crap-detect rumors. If the rule of thumb for attention literacy is to pay attention to your intention, then the heuristic for crap detection is to make skepticism your default. Don't refuse to believe; refuse to start out believing. Continue to pursue your investigation *after* you find an answer. Chase the story rather than just accepting the first evidence you encounter.

The first time I saw my daughter use a search engine to research homework, I explained that "in the olden days, you gathered information by going to the library for a book or magazine article. You might disagree with a library book, but you could be somewhat confident that someone checked the author's claims about facts before the book was published. When you get results from a Web search engine and click on a link, you can't be sure that what you get is accurate or inaccurate information, misinformation, or totally bogus."

"Who can tell me which is which?" she asked, getting to the heart of the Internet's challenge to the age-old authority of texts.

"I'm afraid that's now up to the person who uses information they find on the Web, not the people who put information onto it," I answered, much to my daughter's dismay. "You can get fooled into believing all kinds of wrong things if you don't know how to tell the difference between the good and bad stuff."

To show her what I meant, I typed in the name of the civil rights leader Martin Luther King Jr. I knew that near the top of the first page of results from most search engines is a link to a site titled "Martin Luther King Jr.: A True Historical Examination."[1] It doesn't take long to see that this "true historical examination" presents King as a disreputable character. I asked: "How do you know if this is true or not?"

"It looks like most everything on the Web," my daughter replied.

"Look for an author," I prompted.

She did a search on several of the authors' names whose work was republished on the site and noticed that search result snippets frequently used the word "racist." I showed her how a free Internet service called Whois reveals the name of the registered owner of most Web sites. The true historical examination site turned out to be owned by an organization called Stormfront. A few additional keystrokes revealed that Stormfront is a white supremacist organization. (Years after that discussion with my daughter, Stormfront started publishing "Hosted by Stormfront" on its home page.)

Ever since I started demonstrating to my daughter and, more recently, my university students that all is not what it seems to be online, I've collected URLs for Web sites that appear to be real, but are either willfully misleading like the "cloaked" Web site I showed my daughter or reveal themselves to be hoaxes if you look closely enough. The harmlessness of Web tricks varies. A site that urges readers to support the campaign to "help save the endangered tree octopus" is mildly amusing (and contains clues to its fake nature if you dig deeper), but the site that claims to be an online pregnancy detector is more frightening ("fill in your name and press the 'start test' button").[2,3] (The pregnancy test site disappeared, months after I first found it, although I located it by using Google's "cached copy" service—and in the future I'll be able to use the Internet Archive's Wayback Machine to get a snapshot of what the site looked like before it was removed. Publishing is permanent now; in chapter 6, I'll get into what that means for all of us.)[4] Genochoice.com comes across as highly professional and offers totally fictional capabilities such as cloning yourself.[5] Hetracil.com looks like a slick ad by a pharmaceutical company, except there is no such thing as an "anti-effeminacy drug."[6] I collect links to hoax sites under the tag "Crap Detection."[7]

When my daughter asked, "How can I tell if *anything* I find on the Web is real?" I told her to "think skeptically, look for an author, and then see what others say about the author." In recent years, journalist John McManus has written an excellent guide to crap-detecting journalism, "Detecting Bull: How to Identify Bias and Junk Journalism in Print, Broadcast, and on the

Wild Web." "Think like a detective," advises McManus.[8] Like smart Web searching, good credibility testing is a process, not a one-shot answer. Simply deciding to perform more than one search query, use more than one search engine, or look beyond the first page of search results is the first step in tuning your critical information consumption skills, just as observing your own attention is the first step in learning mindful media use.

My daughter and search engine technology both came of age in the late 1990s. Today, as it was then, the root question for assessing the credibility of a Web page remains, Who is the author? Immediately after that, ask, What are the author's sources? The first source we tested for martinlutherking.org began to raise warning flags regarding possible bias. A lack of sources is as suspicious as sources with known bias. Learn to use Easywhois or another Whois service to find out who owns a site if there is no author listed; use alexa.com to find out approximately how much traffic a Web site receives.[9, 10] Once you know a URL, try pasting it into one of the tools on Network-tools.com.[11] If the author provides a way to communicate or add comments, turn up the credibility meter. If the author responds to comments, read those responses. Is the site a .gov or .edu? If so, increase your estimation of a site's credibility.

Take the Web site's design into account, but don't count on it. Professional design should not be seen as a certain indicator of accurate content (Genochoice and Hetracil are beautifully designed), yet visibly amateurish design is sometimes a signal that the "Institute of Such-and-Such" might be a lone crackpot. Treat a site's design not as validation of credibility but instead as one possible clue (along with grammatical errors, suspicious sources or lack thereof, and other people's negative opinions of the site) that could convince you to lower your evaluation of the site's credibility. If the author provides sources, search the authors' names. Have other Web sites linked to this page, and if so, who are the linkers? Use the search term "link: http:// . . . " (with your URL in place of the ellipses) to see every link to a specified page. A veteran ink-stained-turned-digital journalist I've known since he left the *San Francisco Examiner* to become a founding editor of *Salon*, Scott Rosenberg, has published an excellent guide for consumers of both news and any kind of information online: "In the Context of Web Context: How to Check Out Any Web Page."[12]

All the mechanics of doing this kind of checking take only a few seconds of clicking, copying and pasting, searching, and judging for oneself. The part that requires the most work is learning how to judge.

Journalists talk about "triangulating" by checking three different, credible sources. In January 2011, I saw a report on Twitter that Egypt had cut its country off from the Internet. I didn't know the person who had tweeted this alarming report—the kind that journalist Dan Gillmor calls "interesting, if true." So I looked on the CNN, Al Jazeera, ABC, and BBC Web sites. Nothing. At this point, I not only felt that I should be skeptical but also warned others via Twitter. Several people replied to me immediately with links to independent sources. But three of those sources linked back to the same blog post that had been made a few minutes prior by the author I don't know. Then someone else told me via Twitter that a person I know to be a reliable source was on the phone with a journalist in Cairo when the Internet blackout started happening. So I looked at the reliable source's own tweet stream and verified the report about their call to the Cairo reporter. I had one point of my triangle. Someone else on Twitter reminded me to use "traceroute," an Internet service that reports on the availability of Web sites. Sure enough, there seemed to be an outage in Egypt; the whole country indeed seemed to be unavailable from my part of the Internet. I'm not technical enough to know whether this can be faked or is a frequent temporary occurrence, but it was strong enough to be my second point. When the Associated Press published a report of curtailed Internet access in Egypt, about an hour after I saw the first tweet, I had my triangle and started retweeting my sources.

This told me that Twitter can be an hour or more faster than existing news networks—if you know how to triangulate. Those who failed to triangulate in 2009 about previous reports that "American Airlines will fly medical help to Haiti if you text this number," however, ended up repeating a hoax. The "Gay Girl in Damascus" whose blog was widely followed during the Syrian revolt of 2011 was exposed as a U.S. man (a classic case history of journalistic crap detection).[13]

You aren't paranoid if you suspect that some sites might even deliberately try to deceive you. Political pranksters created Gatt.org, and once duped the Center for International Legal Studies into believing this was the World Trade Organization's Web site. A few years ago, the aforementioned center arranged for someone from the parody site to speak at its annual conference. The speaker—an imposter from an activist group known as the Yes Men—made outrageously racist remarks, claiming to be an authentic World Trade Organization spokesperson. In a staged incident, the speaker was pied and then faked his own death. Disinformation can be well crafted. Look around mercuryfacts.org and see if you can figure out on your own that the site has a pro-fishing-industry agenda.[14] I searched "who is behind mercuryfacts.org," and the top result was a Wikipedia page about a public relations professional who was reportedly hired by the tobacco industry and others to create misleading Web sites.[15] A search for the name of the alleged public relations pro's organization, "Center for Consumer Freedom," led me to Sourcewatch.org, a kind of Wikipedia for people who want to track down front groups online.[16] A good question to ask yourself, particularly when a Web site asks you to download something to your own computer, is, Might somebody be trying to put one over on me?

In recent years, as so many more people have started to rely on the Web for such vitally important forms of information as news, scholarly research, and medical as well as investment advice, the lack of general education in critical consumption of resources found online has become a public danger. No, Gates won't send you five dollars for forwarding this chain email, the medical advice you get in a chat room isn't necessarily better than what your doctor tells you, and the widow of the deceased African dictator is definitely not going to transfer millions into your bank account. That scurrilous rumor about the political candidate that never makes the mainstream media, but circulates as email and blog posts probably isn't true. The data you are pasting into your memo or term paper may well be totally fabricated—and detectable by specialized plagiarism-finding search tools. If nothing else, check Snopes.com, the site that has been debunking online rumors since 1995, before you pass along the latest social media meme infection.

In the previous chapter, I mentioned Stanford professor Fogg in regard to attention training because he's an expert in the use of social media persuasion, self-persuasion, and habit changing. (I often cite the names of people who make scientific claims so my readers can search on my sources, and then judge their credibility for themselves.) Fogg also coauthored one of the earliest studies of how people test the credibility of Web pages. In "The Elements of Computer Credibility," Fogg and coauthor Hsiang Tseng stress that credibility is always a perceived quality, and not a property that can be found in any human or computer product.[17] Think of credibility as a measure of the degree to which you believe the information you are evaluating is accurate. In 2001, Fogg and several other colleagues studied fourteen hundred participants in the United States and Europe, noting that online credibility judgments are affected by references and links to real-world means of verification (brand name, contact information, credentials, and feedback by site users, like Amazon reviews), usability and quality of site design, errors and technicalities such as broken links and frequent site unavailability, and currency and reliability (recency of updates, consistency, and verifiability).[18]

Credentials and highly regarded brands reduce the burden of investigating the credibility of information online, but do not remove it; hoaxes at such distinguished institutions as Harvard and the *New York Times* signal that although they have proven reliable, even the most well-credentialed sources are not 100 percent authoritative.[19, 20] Anybody who has spent sufficient hours or years surfing the Web can sense when a Web site is broken or obviously badly designed, and most people know how to discount sites that have not been updated in ages. Fogg and his colleagues seem to be confirming common sense. A different study, however, uncovered yet another commonly used method that could be dangerously misleading: too many people seem to regard search engines as authoritative in themselves. A study in 2010 by Eszter Hargittai, Lindsay Fullerton, Ericka Menchen-Trevino, and Kristin Yates Thomas, found that students use search engines as a parameter of trustworthiness. As long as a site is toward the top of a search engine's listings, many of this study's subjects considered it credible.[21]

People who are more experienced at Web surfing, some research suggests, do triangulate by applying several methods for assessing credibility. A study by R. David Lankes showed that experienced Web users combine clues (thinking like a detective) from search engines, automated recommendations, aggregation services, and "network recommendations" via social media. Lankes explains that "whereas credibility has historically been tied to concepts of authority and hierarchy, in the reliability approach users determine credibility by synthesizing multiple sources of credibility judgements. Both the need for synthesis and the richer set of resources to be synthesized are products of the pressure for participation enabled and imposed by networked digital media."[22] Lindsay Pettingill's Harvard study found that

in order to determine the presence of these "credibility" criteria,' students used various cues: popularity of site, professional reputation/offline reputation (including professional staff with resources to gather definitive and credible information), previous personal experience with site, proof of "neutral" (not for-profit) affiliation (.gov for government site, .edu for education, .org for non-profits), tone of the writing (neutral vs. opinionated), and elements of style (use of quotes, pictures, by-lines, newspaper layout).[23]

Most people ask themselves whether a detective-like inquiry to verify the answer to a Web search is worth the time. According to researchers Soo Young Rieh and Brian Hilligoss, interviews with twenty-four college students revealed that they would be willing to compromise certainty about credibility for speed and convenience.[24] That trade-off is familiar to older Web users as well. I know that I apply the same calculus, weighing what is at stake before allocating more time. As Shirky puts it, "There's a spectrum of authority from 'Good enough to settle a bar bet' to 'Evidence to include in a dissertation defense.'"[25] Most of the time, trading certainty for convenience is a reasonable decision—but not always. That's where asking yourself whether you have time for another query and considering what's at stake is a useful strategy.

Miriam Metzger, Andrew Flanagin, and Ryan Medders conducted one of the most thorough studies of exactly how people test the credibility of Web pages. They analyzed focus group data from 109 participants with diverse demographic characteristics and different levels of Internet expertise. Noting (as I did to my daughter) that "digital media and information abundance may complicate people's confidence in and knowledge of who is in authority," Metzger and her colleagues added, "Electronic networks and social computing applications make it easier for individuals to harness collective intelligence to help them assess and evaluate information and sources online."[26] I'll soon drill down on that "collective intelligence" aspect of credibility testing. The social aspects of critical evaluation can be powerfully useful, but they also can be misleading. Skill at evaluating the quality of collective intelligence is essential to knowing how to take advantage of it.

Before taking on a more detailed examination of social factors in credibility testing (what Metzger and her colleagues mean when they talk about harnessing "collective intelligence to help them assess and evaluate"), consider that a kind of popularity measurement is critical to the way all search engines work today. Google's famous "PageRank" algorithm (named after Google cofounder Page, who created it as a graduate student) arranges the results of a Web search by measuring how many other Web pages link to every Web page identified in a search. Of all the Web pages that have the words specified in the search query, those with the highest number of inbound links coming to them from other pages are ranked higher in the search results, and links from sites that are themselves linked to by a large number of sites count more than those from less popular ones.

The secret sauce for Google's search is more complicated than that, but counting links as proxy votes for the usefulness of a Web site is fundamental. PageRank takes into consideration the ranking of each page that casts a vote for another page's credibility and usefulness by linking to it — links from some pages are given higher value in the PageRank evaluation." The various flavors of voting on which Google's "+1," Facebook's "like," and Digg's "digg" and "bury" are based can be powerful collective judgment aggregators, but the presence of spam blogs and "content farms" that publish random words or text pasted from other blogs plus the names of brands the spammers want to push are evidence that it can be lucrative to game "Google juice." The entire profession of search engine optimization involves the craft of manipulating the opinion of search engines, just as public relations involves the craft of manipulating public opinion.

Some recent ethnographic research by others indicates that seventh graders who happen to be hard-core gamers, bloggers, or YouTubers are more likely less trusting, and more inclined to look beyond the surface of a Web search—probably because enthusiastic young creators of online content use search and social media to learn their subculture's craft.[27] In my interview with Ito, who directed the most extensive ethnographic study of youth and digital media yet conducted, she noted that "one of the side effects or what we call collateral learning for kids who do engage in geeked-out, interest-driven activities is that when you start engaging in knowledge or media production, you tend to develop a much more sophisticated understanding of how knowledge and media is produced more generally"[28] Metzger, Flanagin, and Medders report that

those who use the Internet to immerse themselves in virtual worlds more often (including playing games such as World of Warcraft) and those who contribute information online, show higher levels of concern about credibility. These results indicate that as kids engage more, and more deeply, with various aspects of the Internet, they may develop a healthy sense of skepticism and concern about the believability of information online. . . . [U]sing more methods to evaluate credibility and being more meticulous in evaluating information leads kids to be more cautious about trusting strangers.[29]

So while age can be a factor in crap-detection fluency, experience and engagement may be more important. A ten-year-old online game enthusiast or videoblogger may do more sophisticated credibility testing than an eighteen-year-old college student who doesn't use the Web much.

Before you can test information, you need to know how to find it. I'll start with search skills, and then look at mental, digital, and social tools you can use to lead a more crap-free life online.

How to Search and How to Know

Looking around online and searching is an important first step to gathering information about a new and unfamiliar area. Although many of these forays do not necessarily result in long-term engagement, youth do use this initial base of knowledge as a stepping-stone to deeper social and practical engagement with a new area of interest. Online sites, forums, and search engines augment existing information resources by lowering the barriers to looking around in ways that do not require specialized knowledge to begin. Looking around online and fortuitous searching can be a self-directed activity that provides young people with a sense of agency, often exhibited in a discourse that they are "self-taught" as a result of engaging in these strategies. The autonomy to pursue topics of personal interest through random searching and messing around generally assists and encourages young people to take greater ownership of their learning processes.

—Mizuko Ito, *Hanging Out, Messing Around, and Geeking Out*, 2009

Search engines are such powerful magic that we've forgotten how magical they really are. While people stand in line for hours to pay for the privilege of walking around a fake village full of actors posing as magicians in the Wizarding World of Harry Potter theme park, millions of people use computer and telephone keyboards to utter magical spells—with various degrees of proficiency and success—every day. If you can cast your search query, the exact words you submit to a search engine, in precisely the right terms, screens full of up-to-date knowledge in multiple media appear before your eyes, literally out of thin air. Your Internet provider charges you for transmitting the data back and forth, but the knowledge comes to you free of charge. What could be more magical than that?

Google, Bing, Yahoo! and other search engines offer search as a free service on the Web, because searches provide the marketing information that advertisers have sought like the conquistadores sought El Dorado—a way to show large numbers of individuals advertisements that each person might actually be interested in. Search is both a public good—something useful to everybody, but that individuals lack sufficient incentive to create for themselves—and a way to amass significant private wealth by selling a valuable commodity. In Web search, the valuable commodity is the searchers' attention. Search engines sell sponsored links that appear on the top or side of the page of links displayed in response to a search query. Whenever someone clicks on a sponsored link, a small amount of money goes to the search engine provider. Those clicks add up to billions of dollars each year.

Access to attention is not just lucrative. It can be political. When petroleum company BP's Deepwater Horizon offshore platform started spewing crude oil into the Gulf of Mexico in 2010, BP bought sponsored link space from major search engines for terms such as "oil spill," "BP's oil spill," and "Deepwater Horizon" in order to provide its side of the story. I bring in Harry Potter and BP to underscore the way that the knowledge magic of search terms depends on how well the searcher phrases the spell. Observations of how people search online reveal that few individuals apply even the most elementary of advanced search operators such as using quotation marks around exact phrases, or using addition or subtraction symbols to include or omit certain words. Google, Bing, and other search engines have a link to "advanced search" on their home page that enables diligent searchers to look for pages on which all the words in the search query appear in any order, exact wording or phrases appear, and one or more of a list of words appears, eliminating pages on which specific unwanted words appear. Knowing how to use advanced search can empower any searcher, but these technical competencies don't necessarily inform the searcher about how to find the information they want in an adequate context. What steps must be taken to turn searching and finding into learning and knowing?

Chris Heuer teaches people to search through his social media club, and offers the good advice to "write the answer you want to get" when formulating your search query.[30] For instance, I wanted to know how and why Thomas Edison (who championed delivering electricity through direct current), rather than his equally brilliant rival Nicola Tesla (who championed alternating current), ended up winning the battle for commercial control over electricity. The answer I wanted to get would complete the sentence "Tesla lost to Edison because . . . " so I entered that phrase (minus the three dots) into Google and Bing. Google's Search Basics puts it this way:

A search engine is not a human, it is a program that matches the words you give to pages on the web. Use the words that are most likely to appear on the page. For example, instead of saying [my head hurts], say [headache], because that's the term a medical page will use. The query [in what country are bats considered an omen of good luck?] is very clear to a person, but the document that gives the answer may not have those words. Instead, use the query [bats are considered good luck in] or even just [bats good luck], because that is probably what the right page will say.[31]

Although I had to know how to follow the trail and assess the quality of the sources, the top page of links for each of the two major search engines provided everything I needed to know and much more. At that point, my infotention training kicked in to remind me that I was writing a blog post today, and needed to stop following the trails of links about electric inventions, genius, and the turn of the twentieth century, fascinating as they might be. On another day, without a deadline, I'd call such unplanned trail following a form of exploratory learning. When investigating a topic, I often add the words "critique" or "criticism" to my search query in order to find contrary or skeptical opinions.

I visited the Googleplex, as googlers call their futuristic headquarters in Mountain View, California, on several occasions to talk with Dan Russell, Google's "search anthropologist."[32] Russell also traveled to my corner of the San Francisco Bay Area to walk in an oak forest and talk about how people search online. One of the first things Russell noted was that sometimes using more words in a search query can help you zero in on the answer, but sometimes making your query too precise "takes you down the wrong rabbit hole." Adding words can return more precise answers, yet that means restricting the scope of the results. Frequently, at the beginning of a series of searches, you want to start more broadly. "Choose keywords that you think will appear on the page you seek, put yourself in the mind-set of the author," Russell suggested.[33]

Russell also advised familiarizing yourself with the anatomy of search page results: the sponsored links are clearly indicated by all reputable search engines, and are most often presented both at the top and in the right-hand margin of search results pages. When you are seeking information about something to buy or somewhere to go on vacation, sponsored links might turn out to be useful. But if you are not interested in information provided by someone trying to sell you something, the other links on a page along with pages deeper in the search engine results are important. The top line of each result, usually presented in bold type, is the Web page's title as a clickable link that will bring you to that page. The short (usually two-line) description below the title is known as the "snippet," and below that is the Web address of the site listed (aka the URL). Sometimes, scanning the snippets reveals a phrase or set of words that would make for a better query—or tell you all you need to know about the topic.

Search, Russell reminded me, can be a gateway to self-learning—which in turn reminded me of Ito's discovery that for young digital culture enthusiasts, searching about a subject is part of a cultural learning process. Ito concluded that youth who "mess around" online by searching for clues to a subcultural topic such as role-playing games, animated cartoon videos, the Pokémon card game, or fan fiction that might seem trivial to others are partaking of a complex social system that uses knowledge exchange for competitive bonding as well as builds learning communities around the tools of their subculture's digital craft. The youths interviewed for Ito's studies had discovered something crucial to all who use Web search: search is most productively used for learning rather than just for finding.

If you don't know much about a topic and seek guidance in learning about it, Russell suggested using search terms such as "tutorial," "introduction," "lesson," "background" in your query. Although educators warn students not to use Wikipedia as a final source or citation, Russell explained that Wikipedia is often a good place to begin an inquiry about a topic. If you can narrow your search down to a word like "metacognition" or phrase like "evolution of cooperation," Wikipedia is likely to include a set of external links that can be excellent starting points. Russell called these starting points "gateway sites," and also recommended cia.gov for international facts and data, reference.com, and about.com (now owned by the *New York Times*). When you are in beginner learning mode, Russell said to include "how-to" or "DIY" in your search to find gateway sites about a topic. I've discovered that going to YouTube and searching on how to do whatever it is you want to learn is more likely than not going to lead you to a detailed video that someone made about doing exactly what you want to learn to do.

I also spent time with a search expert for Google's most recent competitor, Microsoft's Bing search engine. When I blogged about crap detection, I was contacted by Betsy Aoki, senior program manager for Bing. She wanted to know if Microsoft could use my blog post in a compendium of critical thinking resources that the Bing team was compiling for teachers and librarians. I replied that I would grant permission as long as Microsoft didn't try to own it. That wouldn't be a problem, she assured me. Then I asked why she was particularly interested in critical thinking. It turned out that Aoki, a former journalist, was at the forefront of Microsoft's effort to position Bing as a "decision engine." Teaching people to use search and other tools to critically examine the information they find on the Internet was part of Microsoft's strategy. For that reason, Aoki enlisted me along with others in compiling critical thinking as well as online credibility-testing resources for educators and librarians. When I was unable to physically attend a scheduled workshop with teachers and librarians, Aoki helped me create my first videos for educators about crap detection.

Aoki reminded me to look at the snippet for contextual clues (the snippet for Hetracil on both Bing and Google contains the words "[contains fictitious information]"), examine the second, third, and later pages of search results, and use more than one search engine. Although I had not told her that Google's Russell had said the same thing, Aoki also approved of Wikipedia as a site that is, more often than not, a useful jumping-off point, a gateway to a series of searches organized as a learning journey rather than a treasure hunt.

When Aoki first contacted me, she asked, "How can I help you get the message out about crap detection and critical thinking?" I suggested that we enlist teachers and librarians in compiling good resources for teaching critical thinking. Microsoft sponsored me as a speaker at the International Society for Technology and Education (ISTE), a conference of exactly the people I wanted to learn from and reach.[34] During my talk, I launched a wiki, hosted on ISTE's Web site, and invited educators to add to it—an ongoing online resource repository for tools, lesson plans, and hoax sites.[35]

The free e-book that Microsoft has made available as a result of Aoki's work, including my contributions and many others, together with detailed online lesson plans, is a valuable guide for educators who seek specific ways to teach critical thinking.[36,37] I'm happy that Microsoft sees a competitive advantage in encouraging critical thinking and will be happy if Google starts making similar resources available.

My conversations with Google's Russell and Bing's Aoki convinced me that if the two biggest search engines were interested in search literacy, I was on the right track with my nascent ideas about helping people learn to be more critical information consumers. And I confirmed my experience from my own PLN that we already have a well-understood social role for a critical thinking tutor: librarians.

Once you've searched, you need to determine how much you should trust the information your search has yielded. That's where your well-tuned internal crap detector comes in handy.

Tuning Your Crap Detector

So you see, when it comes right down to it, crap-detection is something one does when he starts to become a certain type of person. Sensitivity to the phony uses of language requires, to some extent, knowledge of how to ask questions, how to validate answers, and certainly, how to assess meanings.

I said at the beginning that I thought there is nothing more important than for kids to learn how to identify fake communication. You, therefore, probably assume that I know something about how to achieve this. Well, I don't. At least not very much. I know that our present curricula do not even touch on the matter. Neither do our present methods of training teachers. I am not even sure that classrooms and schools can be reformed enough so that critical and lively people can be nurtured there.

Nonetheless, I persist in believing that it is not beyond your profession to invent ways to educate youth along these lines. [Because] there is no more precious environment than our language environment. And even if you know you will be dead soon, that's worth protecting.

—Neil Postman, *Bullshit and the Art of Crap Detection*, 1969

Use the following methods and tools to protect yourself from toxic bad info. Use them and then pass them along to others. Promote the notion that more info literacy is a practical answer to the growing info pollution. Be the change you want to see.

Although the Web undermines authority (by enabling anybody to publish), authority is still useful as one clue to credibility in a detective hunt that accounts for many other clues. Claims to authority, however, need to be questioned. I might add credibility to my assessment if a source is a verified professor at a known institution of higher learning, an authentic MD or PhD, but I would not subtract it from people without credentials whose expertise seems authentic. If you are going to grant credibility to people whose expertise is based on being a professor of something, make sure that assertion is accurate. Don't stop at simply verifying that the claim to be a professor is valid if you are looking for scientific credibility. The next step is to use the Faculty Scholarly Productivity Index that derives a score from the scholar's publications, citations by other scholars, grants, honors, and awards.[38] If you want to get even more serious, download a free copy of Publish or Perish software, which analyzes scientific citations from Google Scholar according to multiple criteria.[39] Or use the h-index to calculate how many times other scientists have cited a particular source.[40] Again, don't trust just one source; triangulate.

Think of tools such as search engines, the productivity index, and hoax debunking sites as forensic instruments like Sherlock Holmes's magnifying glass or the crime scene investigator's fingerprint kit. For people who bet their health on online medical information, their economic well-being on online financial gossip, or their political liberty on the rumored news they get from Twitter, blogs, or YouTube, the stakes in this detective game are high. For example, you could triangulate by googling the author's name, entering the author's name in the Faculty Scholarly Productivity Index, and using the literacy resources at FactCheckED.org to triangulate a source.[41] FactCheckED.org's sister site, FactCheck.org, researches claims by all political factions.[42] How much stronger would democracies be if citizens checked the political section of the *New York Times'* "About Urban Legends" site or the U.S. Department of State's "Conspiracy Theories and Misinformation" site before passing along a link or an email about, respectively, a political figure or conspiracy theory?[43, 44]

Crap-detection skills and the lack of them are a life-and-death matter for more people every day. The good news about the pace of medical research is also the bad news: few medical specialists can keep up with the rate of new discoveries. That means that it's possible for the collective intelligence of a committed community—and there is nothing as committed as people who are suffering from a disease—to stay ahead of all but the most dedicated individual specialists. Nevertheless, along with the latest word on cutting-edge drug trials are unsubstantiated claims, rumors, and outright quackery. Well-intentioned yet dangerously misinformed people, quacks who sincerely believe that their ineffective cures will save the world, and straight-out charlatans who unblushingly fleece the ailing abound online. It's not just that uninformed consumers of bad medical information can harm themselves; people who link and forward without checking closely are part of the problem. When it comes to medical information, just as when it comes to information that affects political liberty, believing or forwarding bad info can be unhealthy or fatal.

How much work is it to check three links before believing or passing along online health information? Simply googling the name of the person who tried to sell do-it-yourself eye surgery kits, for example, immediately raises questions for those who are considering aiming lasers at their own retinas.[45, 46] Patients who want to learn more about their disease and treatment are not totally at the mercy of the oceans of rubbish. Tools for navigating research reports and treatment options exist. For scientific articles, ScienceDirect has guest access.[47] The Health on the Net Foundation has been a steady source of finding reliable, credible health information online.[48] It even has a browser plug-in that enables you to check health information on any Web site against Health on the Net's database.[49] An astute medical student wrote a quality-check guide to medical information online.[50] The Medical Library Association published "A User's Guide to Finding and Evaluating Health Information on the Web."[51] Start with these gateways if you are new to seeking online medical information.

What person *doesn't* search online about their disease after they are diagnosed? According to the Pew Internet and American Life Project, "[Sixty-six percent] of Internet users look online for information about a specific disease or medical problem."[52] In a *Time* magazine article, Zachary F. Meisel, an emergency physician and clinical scholar, describes the situation:

To debate whether patients should or should not Google their symptoms (which a surprising number of doctors seem to enjoy engaging in) is an absurd exercise. Patients already are doing it, it is now a fact of normal patient behavior, and it will only increase as Internet technology becomes ever more ubiquitous. The average Joe has more health information at his fingertips—both credible and charlatan—than all the medical libraries ever built put together. So the real question is, What can professionals do to translate this phenomenon into better health for their patients and the public?[53]

Meisel suggests that health care professionals encourage their patients to educate themselves about their diseases, and "guide their patients to Internet sites that exclusively present current, peer-reviewed and evidence-based health information."[54] I'm cheered that in an international newsmagazine, a medical professional has publicly advised doctors to teach their patients the kind of crap detection that licensed practitioners learn to do early in their medical careers. Meisel, for instance, points to the Agency for Health-care Research and Quality Web site that provides specific guidebooks for different diseases, directing patients with particular diseases to officially vetted research reports that compare different health care interventions.[55]

The Web site Search Engine Watch, an industry source I consider knowl-edgeable, published a good article by Dean Stephens titled "Turning to Social Media and Search Engines for Smart Health Answers."[56] In regard to getting answers, Stephens recommends MedHelp.org and JustAnswer.com for "detailed information specific to your question from health profes-sionals," and favorably mentions Sharecare.com and Healthline.com.[57, 58, 59, 60] The Pew Internet and American Life Project claims that while medical professionals and Web searches are sources of some types of information for cancer patients, "when it comes to practical advice for coping with day-to-day health situations, people are as likely to turn to peers as they are to professionals."[61]

Today, large professional operations such as PatientsLikeMe.com and CureTogether.com host peer communities for patients as well as their families and caregivers.[62, 63] PatientsLikeMe makes its money by selling anonymized data about patients to pharmaceutical companies—and it is abundantly open about it. In 2011, PatientsLikeMe used its data to crowd-source (more on that later) and publish its own clinical trial questioning the effectiveness of using lithium carbonate to slow the progress of the disease amyotrophic lateral sclerosis. By comparing patients who took lith-ium with similar patients who did not, the online social network presented evidence strong enough to stimulate more rigorous studies. Although not a gold standard double-blind trial, this patient-initiated research is an early indicator.[64] CureTogether encourages "patient-driven research," and relies on funding by its founders and angel investors. OrganizedWisdom.com curates and organizes information from thousands of health and wellness experts.[65] TalkAboutHealth.com matches patients with peers and experts—a kind of matchmaker for support groups.[66]

With the resources provided in this chapter and a willingness to regard medical Web searching as a learning journey rather than a race to find an answer, informed patients and caregivers have more than enough tools to find good information, but also detect the bad information related to any medical condition or treatment.

If there is anything more important than health, it's liberty. As James Madison put it, "A popular Government, without popular information, or the means of acquiring it, is but a Prologue to a Farce or a Tragedy; or, perhaps both. Knowledge will forever govern ignorance: And a people who mean to be their own Governors, must arm themselves with the power which knowledge gives."[67] Madison was pointing at the significance of relatively accurate, unbiased, uncorrupted journalism to the functioning of a democracy. Digital media and networks have challenged the institutions and practices of journalism just as they have undermined the authority of medical professionals as the sole source of valid information about diseases. In the medical field, however, doctors and nurses are still the best caregivers for the ailing. In journalism, two billion Internet users, more than five billion cell phones, hundreds of millions of bloggers, Twitter and Facebook users, and hundreds of millions of camera phones have broadened as well as democratized the field, and at the same time multiplied the need for critical news consumers.

The power of nonprofessionals first came to light in the 1990s, when bloggers on the Left independently researched a story that mainstream media appeared to have let fade: the history of racist remarks by U.S. Senate majority leader Trent Lott, who was forced to resign his leadership post.[68] As the power and spread of social media grew, acts of citizen journalism began to grow more common. The first pictures of the London terrorist bombings of July 7, 2005, for example, came from camera phones at the site of the attack.[69] Along the same lines, the first pictures from the site of the 747 that landed in the Hudson River in 2009 came via smart phone and Twitter.[70] And smart phones sent video directly from the streets of Tehran in June 2009 and Cairo in 2011.[71]

Before Twitter came on the scene, online services like Flickr and YouTube enabled users to tag photographs and video with keywords, making it possible to search for images tagged with those keywords, revealing all the still images and videos coming in from amateur chroniclers during an event. Because Twitter didn't have a native tagging capability, a Twitter user invented the "hashtag" by suggesting that people signal common interests by putting the # symbol in front of words—like #jan25, the hashtag of the Egyptian popular uprising of 2011.[72] The *San Diego Union-Tribune* called publicly for citizen reporters to use the same tags and hashtags for their images of that area's wildfires in 2007.[73] Using the search facility on Flickr or YouTube enables you to see a stream of images or videos, and automatically subscribing to that search through Really Simple Syndication (RSS; explanation to come) means you can continue to see visual reports stream in as others upload them—in real time.

Although traditional gatekeepers have been disintermediated through the widespread adoption of digital tools and networks, the need for trained (if not professional) verification of raw news reports is greater than ever. Anybody can send a video or tweet from the scene of breaking news. When those reports are vetted, situated in a context, coherent stories of real people involved are told, and spokespeople for different points of view about the reports are quoted—all that is when journalism is being committed, whether or not the individuals engaged in the verification, contextualization, and storytelling work for brand-name institutions. Like it or not, journalism is becoming something more akin to a network than a guild.

Many journalists believe that the news cycle accelerated past the breaking point with *Newsweek*'s decision in 1998 to wait for confirmation while Internet reporter Matt Drudge broke the Monica Lewinsky story.[74] The events in Iran in June 2009 made it clear to the world that the increasing velocity of information was overwhelming the traditional public filter of brand-name journalists who staked their careers and their networks' reputation on getting the story right before broadcasting it. Citizen reporters, credible and otherwise, using Twitter and YouTube swamped even the capabilities of previously fast-moving electronic journalists in vans with camera crews. When the political demonstrations happened in 2009, people in Iran and around the world used the now-famous hashtag #iranelection, and for a few days flooded the world with riveting images, shocking and politically inflammatory videos, torrents of contradictory reports, rumors, and apparent disinformation, along with both informed and ignorant political arguments. At one point, 221,000 tweets about Iran passed through Twitter each hour.[75]

As the Iran events unfolded, Marc Ambinder wrote an astute article in the online *Atlantic*, in which he advised, "Follow the developments in Iran like a CIA analyst." Just as thinking like a detective is a strategy for trying to determine the credibility of Web info, thinking like an intelligence analyst is a strategy for trying to gauge the credibility of online reports about breaking news events. Ambinder recommends watching for disinformation, looking for patterns in the geographic location of sources (but warns against assuming that everything that resembles a pattern really is one), examining your assumptions, and seeking out sources that contradict them.[76]

When I started paying attention to citizen journalism, Gillmor, the traditional print journalist mentioned above, was one of my trusted sources. A veteran reporter for the *San Jose Mercury News*, then regarded as the newspaper of record for Silicon Valley, he was one of the first newspaper reporters to take up blogging in the late 1990s. Gillmor didn't stop at blogging. Not only did he permit readers (who he started calling "the former audience") to comment but he replied to their comments too. And he enlisted his online network's help in providing tips and contextualizing the stories he was pursuing. I used Gillmor's book *We the Media* as a text when I taught digital journalism at Stanford University.[77] Gillmor and I have talked often about the need for more effective crap detection, and I'm happy to say that he published a book, *Mediactive*, out of his work at the Berkman Center for Internet and Society at Harvard University and the Walter Cronkite School of Journalism and Mass Communication at Arizona State University. *Mediactive* seeks to improve the public sphere by helping people not only become more savvy news consumers but also encouraging them to be active, mindful participants in news making.[78]

Gillmor's five "Principles of Media Consumption" add up to "core principles for turning mere consumption into active learning." The first principle, "Be Skeptical," is in line with what I've written here so far. The second one, "Exercise Judgment," cautions against reverting to cynicism in response to the unreliability of online information: "Clearly, we need to ask ourselves what kind of society our kids will inherit if they don't trust or believe anyone but their friends, regardless of whether those friends are well informed." Principle number three, "Open Your Mind," is an encouragement to seek out legitimate sources that disagree with your own beliefs.[79]

One of the fears cited by scholars such as Cass Sunstein, head of President Obama's Office of Information and Regulatory Affairs, is the tendency of individuals to pay attention to only those online sources that reinforce their own beliefs.[80] This has been called the "echo chamber" effect, in which bloggers read and quote other bloggers they agree with. In 2011, Eli Pariser, former online organizer for Moveon.org, published his book *The Filter Bubble*, in which he detailed how search engines use precise information about your interests and search history to customize your searches; people who mostly search for and click through to liberal political sources are less likely to see conservative sources, and vice versa—a factor that helps enclose people into bubbles without their conscious knowledge.[81] Getting outside the echo chamber and bubble requires the conscious cultivation of sources beyond your usual ones. For example, if all the news you pay attention to is from the United States, Gillmor recommends GlobalVoices, an online project that aggregates reliable blogs from around the world.[82]

Gillmor's fourth principle, "Keep Asking Questions," is also in line with the "think like a detective" investigative mind-set. Principle five, "Learn Media Techniques," reinforces the learning value of engaging actively in social media as a way of participating in the production of culture and familiarizing yourself with the attitude of the cultural producers whose product you consume.[83]

Some people are exploring the use of social media for crap detection about journalism. FairSpin.org's community votes on stories in order for its aggregate judgments to identify opinion disguised as fact, and reflect the degree of political bias detected in stories from both the Left and Right.[84] NewsTrust.net is an online community of reviewers (around twenty-two thousand as of this writing) who use a set of review tools devised by veteran journalists.[85] I was one of its first members (and served on its board of directors). NewsTrust may or may not represent the way people curate the news for each other in the future, but it's definitely one of the experiments to watch in crowdsourcing critical thinking about journalism.

I can submit any story for review, and NewsTrust's home page then offers a selection of stories already reviewed by other community members. When I opt to review a story for NewsTrust, a menu pops up. I'm asked to rate on a scale of one to five whether the story is factual, fair, and well sourced, and whether (again on a one-to-five scale) I trust the publication or would recommend the story. NewsTrust has one-click-away details on how to assess a news story for factual claims, possible bias, and source credibility. While NewsTrust aims for depth of critical analysis, other applications of social media to journalistic crap detection are trying to do something about the difficulty of detecting credible sources at times of high-velocity reporting—from "social curation" to "crowdsourcing the filter," which I'll discuss later in this chapter.

Bad info isn't the only daily hazard for the mindful digital citizen. There's also the issue of too much information, too quickly. Infotention combines a mind-set with a tool set.

Infotention: Attention to Information

In an information-rich world, the wealth of information means a dearth of something else: a scarcity of whatever it is that information consumes. What information consumes is rather obvious: it consumes the attention of its recipients. Hence a wealth of information creates a poverty of attention and a need to allocate that attention efficiently among the overabundance of information sources that might consume it

—Herbert Simon, "Designing Organizations for an Information-Rich World," 1971

In one window on the screen to my right, I'm composing a PowerPoint presentation about social media and collective action. In one window on the screen to my left, I see that there is new email in my queue. I also note that in another window, one of the people I follow on Twitter has linked to something about a business that runs itself as a wiki. The tweet catches my eye. I ignore my email, switch my attentional focus away from my PowerPoint in progress, and click on the link in my Twitter stream. The link leads me to a five-minute video that's perfect for the presentation I'm composing. I download the video and attach it to my PowerPoint. I also bookmark the video's location with the tags "wiki," "video," and "collaboration." Although I use information technology skills in this brief slice of my life online, including tuning Twitter networks, downloading videos, and social bookmarking, I've also been exercising attentional skills that took me years to learn.

Because I'm a writer, my attention is an evolutionary battleground where potentially enriching or distracting new information feeds as well as competes with the articles and books I need to produce in order to make a living. When do I "get back to work" exactly, and when is my meandering attention precisely what I'm supposed to be doing in order to keep up with my field? For my first ten years as an aspiring writer, my mind wandered often to the wall, the window, the dishes that needed to be done, and the magazine to browse, but eventually, every day, I had to wrench my attention back to the blank page in front of me and try to activate my fingers on the keyboard. Since the early 1980s, however, the digital world along with the network of fascinating but potentially distracting links, videos, tweets, emails, wall posts, and wittily captioned photographs of cute cats has been both the seductive distraction and relevant knowledge flow that have competed for my mind.

The moment you see a link, and then decide whether or not to click it, is the moment you exert executive control of your attention or, by not exerting control, allow it to be captured. When you open your laptop in class, back channel with your BlackBerry in a business meeting, or text while walking down the street or attending a museum with your child, you engage in cognitive and social acts that affect your mind as well as your society. Infotention, as noted earlier, is a word I came up with to describe a mind-machine combination of brainpowered attention skills and computer-powered information filters. It consists of three elements:

• Honing the mental ability to employ the form of attention appropriate for each moment is an essential internal skill for people who want to find, direct, and manage streams of relevant information by using online media knowledgeably. Knowing when not to text, when to bring your attention back on task, and when to take a break from all media is necessary, but not sufficient for successfully navigating the info flow. Not drowning is not the same as swimming.

• Knowing how to put together intelligence dashboards, news radars, and information filters from online tools, like persistent search and RSS, is the external technical component of information literacy.

• Together with and in addition to your own attentional discipline along with use of online power tools, infotention involves sociality—other people. Increasingly, most of the recommendations that make it possible to find fresh, useful signals amid the overwhelming noise of the Internet are transmitted through social media—online networks that mediate social exchange and relationship. Tuning and feeding our PLNs is where the internal and technological meet the social.

Mentally trained, technologically augmented, socially mediated infotention is about continuously detecting information that could be valuable specifically to you, whenever and wherever it is useful to you. Tuning your input doesn't guarantee that what comes in will all be accurate, or that paying obsessive attention to it will be healthy. That's where crap-detection skills and mindfulness come in.

Billions of humans can't deny that they're trying to cope with always-on, available-everywhere, raw data. Yet few people know that the history of information overload goes back centuries, and that we've found ways to deal with it before.

Two contradictory statements seem to be necessary to portray the state of today's info-overloaded cyber-BlackBerrians:

• The flow of information available is accelerating at an unprecedented rate for digitally connected people, from eight-year-olds with smart phones to every adult who lives in fear of an overflowing email in-box.

• At the same time, cries of alarm over cognitive damage caused by the latest communication medium echo ancient fears that seem to cycle over the centuries the way certain email canards recirculate endlessly in the infosphere (please do not send greeting cards to Craig Shergold, the little boy in England with the brain tumor; he's been well for decades, and the postmaster in his village is not happy with the situation).[86]

In August 2010, Google CEO Eric Schmidt dropped a mind-boggling statistic at a high-tech conference: every two days, humans produce as much

information as we did from the era of cave paintings up to 2003.[87] I looked
up a few more stats. SMS messages were up to seven trillion annually by
2011.[88] Facebook displayed a trillion ads in 2010.[89] According to a study
done at the University of California at San Diego, the average American
consumes thirty-four gigabytes of information on an average day.[90] Clearly,
the raw volume and velocity of information as well as opportunity for distraction now is unprecedented. But I find the info-overload fears of the past
to be instructive in the way they eerily reflect today's moral panics about
the putative stupefying effects of the Web, and in the hopeful clue that
history conveys—people responded to overload in the past by developing
mind tools to elevate the information-handling capacities of literate people.

In 1755, French writer and philosopher Denis Diderot warned:

As long as the centuries continue to unfold, the number of books will grow continually, and one can predict that a time will come when it will be almost as difficult to
learn anything from books as from the direct study of the whole universe. It will be
almost as convenient to search for some bit of truth concealed in nature as it will be
to find it hidden away in an immense multitude of bound volumes. When that time
comes, a project, until then neglected because the need for it was not felt, will have
to be undertaken.[91]

That project, for Diderot and his Enlightenment confreres, was the first
encyclopedia. Too much knowledge? Let's try to organize it! Wikipedians
are Diderot's intellectual heirs. Writing of strategies for coping with information in the sixteenth century, Ann Blair notes that Conrad Gesner complained of the "confusing and harmful abundance of books" in 1545, barely
a century after the advent of printing, and that Adrien Baillet warned, a
century after Gesner, that "we have reason to fear that the multitude of
books which grows every day in a prodigious fashion will make the following centuries fall into a state as barbarous as that of the centuries that
followed the fall of the Roman Empire."[92] Alphabetic indexes, reference
books, bibliographies, note taking, authors, and critics were just some of
the tools invented in response to the problems Gesner and Baillet worried
about. And before print, responses to the problems of scribal overload led
to silent reading, punctuation, and the codex form that books still take at
present.

The print overload revolution created problems and stimulated solutions
in the sciences as well. According to Brian W. Ogilvie, the overabundance
of printed descriptions by naturalists overwhelmed "the limited bounds
of factual knowledge that had characterized ancient and medieval natural history, creating an information explosion with which naturalists have

struggled ever since."[93] Biological classification systems were invented to make sense of this overabundance of unorganized data. The history of previous information overload panics and the tools that were devised to cope with them suggests that a period of info overwhelm after the invention of every radically more efficient means of communication tends to be greeted by alarm, followed by new info tools and the growth of newly literate populations. The problem has never been definitively solved. The recurring cycle of inventing cultural methods to filter and organize the plenitude unleashed by new tools of mass publication, thus making possible even-newer publishing tools to create even more abundance, constitute the kind of coevolutionary arms race (like predator and prey) that accelerates innovation in biology.

The tools that are evolving in response to the most recent and biggest info crisis begin at the cognitive level. Trained human attention and judgment is required, or the technical leverage is useless. Before learning how to build radars, filters, and dashboards, it helps to begin by reviewing the basics of mindfulness. Formulate goals and turn them into intentions by paying attention every once in a while to what you are doing at the moment, and then reflecting briefly on how what you are doing relates to your larger goal. That's all you need to get started; when you sit down in front of a screen, write out by hand your goals for each day on a piece of paper, and put it on your desk at the periphery of your vision. A few times each day you will notice the paper. At those moments, ask yourself how your current activity online fits your own goals for the day (and take a deep breath). Practice mindfulness in your use of media in small ways, find places to fit practice into your day, repeat until you've established a new habit of paying some attention to what you are doing with media—texting, gaming, reading, writing, Web surfing, or any one of the 1,001 ways to use your mind online.

If Stanford's Fogg is to be believed, small steps at regular intervals, repeated over time, form a highly effective way to establish new habits. When you are familiar with simply paying attention to your mind on media on a regular basis—daily, hourly, every ten minutes, or whatever you can manage—turn your ever-stronger attentional skill to improving the quality of the microdecisions you make online.

So far, I've been talking about strategies. Microdecisions are tactical; they are what you use to take on information overload minute by minute. Trauma teams perform triage in emergency rooms, on battlefields, and at the scene of natural disasters. They look at each casualty, and decide

whether this person will probably live without immediate attention, are probably going to die no matter what assistance they are given, or are likely to live only if they receive immediate care. Then they focus their efforts on the people in the third category. Attentional triage is not about life and death but instead ultimately about the quality of life for the practitioner.

I've established the habit of asking myself: Is this information or opportunity to communicate worth my attention at all, given my goal for today? Is it something I want to look at later, but not right now while I am focusing on a task toward a larger goal (in which case, maybe I'll open a tab on my browser)? Is it information that I don't want to distract myself with at the moment and don't want to burden my near-term reading agenda, but might want to refer to later, because it is about a subject that interests me (in which case I'll tag and bookmark)? Like basic mindfulness, paying attention to microdecisions—and learning to make them more and more quickly—is easy enough to start, and yields increasing power to diligent, regular practitioners. It's an exercise in strategic goals (What am I setting out to do?), attention (What am I about to click on?), and intention (I'm going to ignore this, or open a tab or bookmark, because I intend to return to focus) that deepens the more you do it.

People and machines publish information so rapidly, and search engines index it so quickly, that we've been living with the "real-time Web" for some time now. I've taught my digital journalism students to use free tools such as RSS aggregators, persistent searches, and engagement filtering services to create dashboards that enable them to swiftly tune into the latest, most credible streams of information on any topic, from a devastating cyclone in Bangladesh to a street demonstration in Cairo, a medical or technological breakthrough, or industry-specific events. The keywords are "tune" and "stream." The real-time Web is not a queue (like your email in-box) but rather a stream. Don't even try to keep up. Learn to sample. You aren't digesting volumes of input. You are tuning your antennae for the right signals and bringing the right information to you while you turn attention away from information that isn't right for you. This partly involves knowing where to put your sensors, and partly how to set up persistent searches.

Start by using the fundamental tool of infotention dashboards: the RSS feed. RSS permits any subscriber to a Web site to be instantly informed whenever that Web site is updated. Web services called RSS aggregators or RSS readers such as Google Reader, Bloglines, Netvibes, or Apple's Mail application allow people to subscribe to blogs or other streams of information enabled by RSS (increasingly, nonhuman sensors like buoys or robot cameras on Mars can send out RSS streams). If you find a site you want to subscribe to, there is an icon or link somewhere on the site to help you copy the short Web address—a kind of URL—that you enter into your aggregator in order to get future updates.

Robin Good, a source I've come to trust in recent years on matters of information literacy, introduced me to the notion of bundling RSS feeds together strategically into what he calls "news radars." A growing number of services make it easy to set up radars if you don't want to roll your own; Good published an excellent guide to these "content curation" tools.[94] I also use paper.li, a service that turns all the URLs tweeted by all the people I follow (or lists of people I designate as knowledgeable about specific topics) into a daily newspaper with headlines, snippets, video, and slides.[95] The layout of newspaper pages is an infotention methodology that has jumped from print to screen-based info flows precisely because it takes advantage of our attentional habits (and trains them). If I know how to find who the experts are, I can compile a daily Twitter-fed news stream about any topic, neatly organized and formatted for my attention. Similarly, aggregage.com puts together custom "newspapers" about specified content.[96] Roll your own feed reader or use templated services, but by all means organize and continuously curate your feeds.

Search and RSS can be combined. Click on "advanced search" on the home page of Google News or Yahoo! News, perform a search, and then subscribe to that search, so that any time one of the tens of thousands of news publications around the world indexed by those services publishes the set of words specified in your search in the future, it will show up in your aggregator. Set up a Google alert, and have it delivered as a "feed."[97] When this book is published, I will know within minutes if the title (or my name) is mentioned in a major newspaper somewhere in the world in any of five languages. Whenever fast-breaking events occur, people often agree on a common hashtag like #jan25 for the 2011 demonstrations in Cairo. I can subscribe to RSS feeds from that hashtag on Twitter, Flickr, and YouTube. You can see how this could be useful to journalists.

I currently use the Netvibes RSS reader because it provides three levels of organization that I can sync with my mental priorities.[98] I have public dashboards, private ones, another one I use for academic matters, and still another for changing interests. Each dashboard makes it possible for me to easily create a series of tabs for different topics. I have a Netvibes "page" for digital journalism, for example, and tabs on that dashboard for new tools, new methods, citizen journalism, crap detection, and the news business. I can drag and drop the tabs to different horizontal positions. Positioning tabs is one way I use visual organization to sync my attention with my dashboard. The cognitive act of categorizing is an important link between digital information-management tools and attention, the way breath connects brain, body, and intention. I use my dashboard as an aid to organizing my attention. The most goal-relevant tabs are on the left, and the least goal-relevant ones are on the right. Some days, I only click on the leftmost one or two tabs to quickly scan the latest feed headlines, and from there, make a microdecision about whether to investigate further at that instant.

At the tab level, I keep the most goal-relevant feeds in the leftmost columns and put the most frequently updated feeds in the top row. Within the feeds themselves—displayed as boxes of adjustable sizes, arrayed in panes on the page—I can scan lists of headlines and snippets from posts. Clicking on a headline opens a pane in the dashboard that shows the blog post or search result. I can browse and scan only the headlines quickly, or click through to the sources and read in greater depth (and at greater switching costs). I have RSS feeds for specific experts' tags in social bookmarking services. I delete feeds when I find them less than useful, and add them when I want to explore a new topic or learn more broadly about it. Within a few minutes of breaking news, it is possible to set up a dashboard with Google News and Yahoo! News searches, Google alerts, hashtag searches on Twitter, blog posts from experts, and feeds on tags from Flickr, YouTube, and Delicious. Without knowing any programming, I can use Yahoo! Pipes to merge all those inputs into a single RSS output that others can subscribe to if I give them the URL. Taking search as a learning instrument to the next level of engagement, dashboards can be used educationally as well as journalistically. I know a sixth-grade teacher who had her students create dashboards instead of writing papers.[99]

Knowing how to ignore the irrelevant is the first step, but knowing how to get the relevant to come to you is the next one. Learning how to find and evaluate good sources of accurate information is a cognitive issue, but we're beginning to see the automation of cream detection—semiautomated, semisocial, infotention amplifiers. Can crap and cream detection both be automated? My former student, Johan Jessen, wrote a masters thesis about how to create an automatic crap detector, and is now seeking funding to start implementing it.

Jessen cites the innovation of WikiTrust, a reputation system for Wikipedia authors and content hosted at the University of California at Santa Cruz.[100] After installing the WikiTrust add-on, users see words highlighted on Wikipedia pages in different colors, depending on whether the text has a high or low reputation. Text that has not been edited recently and/or repeat-edly and editors whose revisions have not been edited for a greater length of time have a higher reputation than text that has never been edited. Jessen also pointed me to Wiki-Watch, a project of the Study- and Research-Cen-tre on Media Law of the European University Viarina Frankfurt, Germany, which assesses Wikipedia articles by checking the number of sources, the number and reputation of editors, and the number of links directed to an article.[101] By compiling and combining different social and algorithmic methods, Jessen hopes to expand such assessment instruments beyond Wikipedia to build "a fully-automated credibility evaluation tool."[102]

Shirky has described the pre-Web era of publishing as working on a "filter, then publish" paradigm, subjecting text to editors and publishers before making it available; now, Shirky observes, the paradigm has flipped to "publish then filter."[103] In that sense, Shirky adds, "there is no such thing as information overload, there is only filter failure."[104] As I've explained, Google's PageRank algorithm is based on the aggregation of popular opin-ions to create a pretty good, if not trustworthy, ranking of Web pages, and other online communities like eBay, Digg, and Facebook aggregate opin-ions for different reasons. Digg is a metanews site where participants don't create original stories but instead vote on whether to "digg" or "bury" news stories and videos listed on their home page.[105] The Digg community is so large that a highly ranked story can crash a small publication's server (this is true of a number of other sites such as Slashdot and 4chan). Such popularity-based aggregated rankings can be gamed. One blogger discov-ered that a clique of ultraconservative participants (the "Digg Patriots") had conspired to bury stories with a more liberal viewpoint.[106]

Shirky labels this emerging system that combines digital aggregation with human opinions "algorithmic authority." He defines algorithmic authority as "the decision to regard as authoritative an unmanaged process of extract-ing value from diverse, untrustworthy sources, without any human stand-ing beside the result saying 'Trust this because you trust me.'"[107]

As an example, Shirky cites a statement that could be right or wrong: "Khotyn is a town in Moldova."

This is a compressed telling, and swerves around many epistemological potholes, such as information that can't be evaluated independently ("I love you"), informa-tion that is correct by definition ("The American Psychiatric Association says there is a mental disorder called psychosis"), or authorities making untestable propositions ("God hates it when you eat shrimp.") Even accepting those limits, though, the as-sertion that Khotyn is in Moldova provides enough of an illustration here, because it's false. Khotyn is in Ukraine.

And this is where authority begins to work its magic. If you told someone who knew better about the Moldovan town of Khotyn, and they asked where you got that incorrect bit of information, you'd have to say "Some guy on the internet said so." See how silly you'd feel?

Now imagine answering that question "Well, Encyclopedia Britannica said so!" You wouldn't be any less wrong, but you'd feel less silly. (Britannica did indeed wrongly assert, for years, that Khotyn was in Moldova, one of a collection of mistakes discovered in 2005 by a boy in London.) Why would you feel less silly getting the same wrong information from Britannica than from me? Because Britannica is an authoritative source.

Authority thus performs a dual function; looking to authorities is a way of increasing the likelihood of being right, and of reducing the penalty for being wrong. An authoritative source isn't just a source you trust; it's a source you and other members of your reference group trust together.[108]

With Google, that reference group is everyone who puts a link on their Web page. With Digg, it's the Digg community. Crap detection must be applied recursively to communities that claim authority, but these communities provide yet another vector to combine with others as inputs to judgment. Both crap detection and knowledge finding have become social. When I interviewed him, Lee Rainie, director of the Pew Internet and American Life Project, noted:

The data have shown us that as the information environment gets more complicated, and as people struggle to keep up with the volume, velocity, and variety of information, they turn to their social networks for a lot of cues about attention and meaning and even audience for their reaction to things. Networks more and more matter to people as they are making decisions or solving problems or trying to get their needs met. I mean literally our data show that people increasingly scan their networks as avidly as they scan the headlines to figure out what the news of the day is.[109]

We're beginning to see raw beams of algorithmic authority filtering through prisms of socially curated judgment. What if, after a process of testing, you decide to trust a particular person or group of people to know something about Moldova or nanotechnology? Twitter makes it easy to create a list and subscribe to ("follow" in Twitter speak) those people's tweets as a group. You can use a Twitter client to switch your stream to only those tweets that come from that group, or more interestingly, you can feed that list to paper.li and get an edited newspaper. What if you take that expert opinion and aggregate it? Sulia.com combines this kind of aggregation (I'll get down to more detail about crowdsourcing in chapter 4) with editorial expertise by treating Twitter lists as a kind of indication of trust.[110] People who put other people in lists named "Egypt" or "search engines" are casting votes for the expertise of the individuals on those lists. Lists can be automatically polled (Sulia claims to crawl millions of them), combined, and then output as what Sulia calls "channels." Sulia claims to add another dimension of human editorial insight by creating editorial networks whose expertise is in recognizing expertise.

Research into combining algorithmic methods and human curation to find the best information as well as eliminating the worst is just getting started. Michael Noll, Ching-man Au Yeung, Nicholas Gibbins, Christoph Meinel, and Nigel Shadbolt presented a paper in 2009 titled "Telling Experts from Spammers: Expertise Ranking in Folksonomies."[111] Noll and his colleagues applied their algorithm to a data set of a half-million users of the social bookmarking service Delicious.com and claim to be able to automatically detect experts, in part by looking for the first people to bookmark a resource that ends up being bookmarked by many other users. A folksonomy is the classification scheme that emerges when large numbers of people apply their own categories (for example, tags) instead of fitting them into a predesigned categorization system (an ontology). Remember how the dictionary, the index, and classification systems developed in response to the print revolution's info overload? It's happening again.

Research turns into product development rapidly in the Web world. Anything that works in a controlled study today might end up on your Web browser within a year or two. One experiment is already beginning to show promise: automated tools for detecting subtle manipulation, recognizing bad rumors, and finding credible news in fast-moving streams. Indiana University researchers have developed a system they call "Truthy" for tracking the emergence and spread of misinformation on Twitter.[112] Fans of the television program the *Colbert Report* will recognize an homage to the July 31, 2006, edition, when Stephen Colbert satirically proposed that a new kind of socially engineered truth had been created by reliance on social knowledge aggregation communities like Wikipedia. Colbert introduced the words "Wikiality" and "truthiness." He suggested that his viewers introduce a spurious statement of fact into Wikipedia, thereby creating a kind of Wikipedia-mediated reality—Wikiality.[113]

Truthy detects rumors, uses epidemiological models to track them to their origin, and enables Web surfers to report suspicious sites by clicking on a "Truthy button." Truthy's research objectives proclaim:

We also plan to use Truthy to detect political smears, astroturfing, misinformation, and other social pollution. While the vast majority of memes arise in a perfectly organic manner, driven by the complex mechanisms of life on the Web, some are engineered by the shady machinery of high-profile congressional campaigns. Truthy uses a sophisticated combination of text and data mining, social network analysis, and complex networks models. To train our algorithms, we leverage crowdsourcing: we rely on users like you to flag injections of forged grass-roots activity. Therefore, click on the Truthy button when you see a suspicious meme![114]

Only time will tell whether Truthy will defeat or make a dent in the spread of scurrilous rumors in future election cycles.

Infotention tools that enable people to tune their own filters are emerging. DataSift.net filters Twitter's seventy million tweets per day, or eight hundred tweets per second, by applying a rules-based engine (for example, "show me tweets containing 'google' from users who don't have 'social media' in their bio, and who have more than five hundred followers," or "show me tweets from my curated Twitter list of tech brands that have more than a hundred retweets").[115] PARC has offered its experimental recommender system "zerozero88" to "address information overload by helping users avoid missing important or relevant content buried within their information streams."[116]

The urgent need by journalists and crisis responders to identify reliable sources of news amid the torrent of tweets that were transmitted during the terrorist attacks of 2010 in Mumbai led to a more directly pragmatic experiment. SwiftRiver is an ongoing attempt to develop sociotechnical systems to crowdsource the filter: by combining the judgment of known experts, social trust measures, and the digital aggregation of mass communications. SwiftRiver aggregates data from many sources, such as Twitter, Flickr, YouTube, and other real-time streams. Text streams can be submitted to natural language processing, and automatically tagged with "who," "what," "where," and other tags; the autotagging can be checked in near real time by communities of volunteers with computers or smart phones (hence, crowdsourcing the filter). The machine-and-human-filtered information stream is then further refined by eliminating duplicates, clustering event reports, and assigning a weighted "reality score." As each new event stream tests the system, algorithms are refined to match those reality scores that ended up being accurate predictors. Filtered data are transformed into real-time, dynamic, animated visualizations that enable networks of experts (journalists and others) to "read" the torrent of event reports more effectively, with more clues to add to their own triangulation methods.[117]

Once you pay attention to attention itself, and practice crap detection and infotention, you're ready to reap maximum benefit from the next literacy: participation. The first two literacies are internal, cognitive, and individual. The last two literacies of collaboration and network smarts are external, social, and collective. Linking them is the act of participation that connects the individual mind to the web of digital culture.

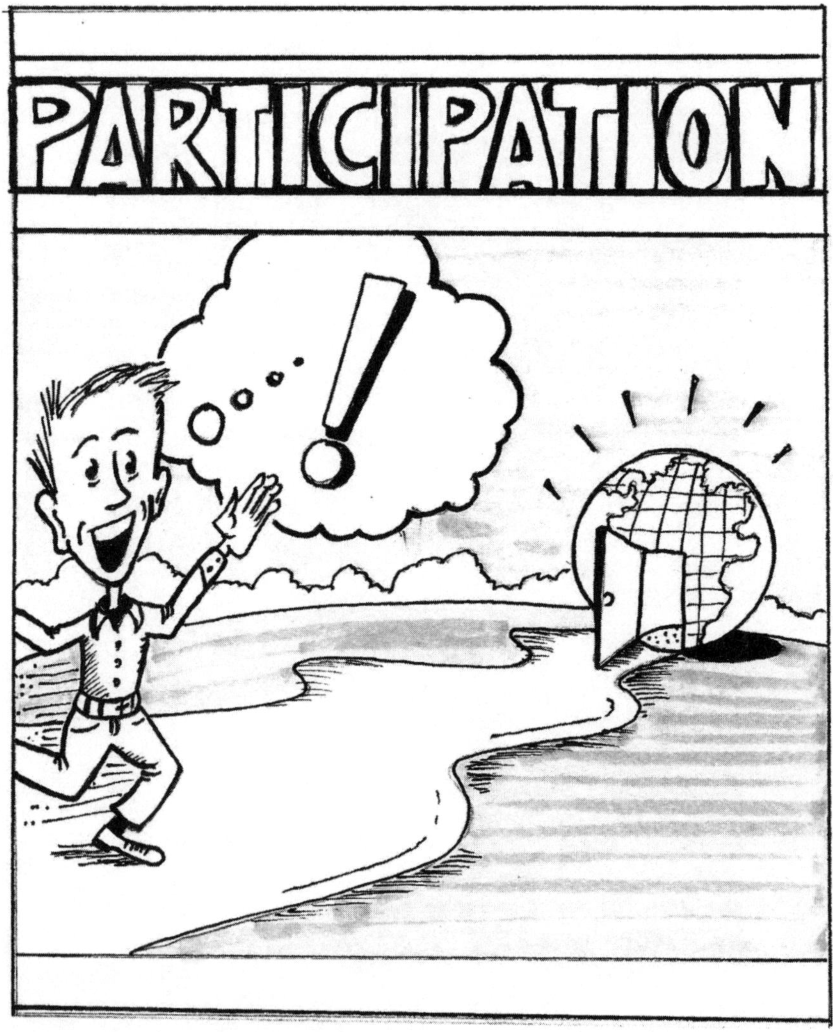
PARTICIPATION

3 Participation Power

Amateur media production is a distinctive cultural domain characterized by an ethic of peer-to-peer appropriation and experimentation that differs fundamentally from more professionalized forms of media production. And this networked culture is fueling a whole host of social, cultural, and technical innovations that are radically reshaping our media landscape.

—Mizuko Ito, "Amateur Media Production in a Networked Ecology," 2010

Consider how the power of digital participation grew during the first decade of the twenty-first century. On February 21, 2001, the attorneys for Warner Bros. surrendered to the group Defense against the Dark Arts, led by Heather Lawver, who had organized online a worldwide boycott to protest the corporation's legal suppression of a Web newspaper created by fans of the Harry Potter books; Lawver, it turned out, was sixteen years old at the time.[1] On July 8, 2003, Bev Harris, a previously obscure U.S. activist, blogged about secret details of the voting machines manufactured by Diebold Election Systems, exposing security weaknesses in a privately owned technology that had become pivotal to the workings of democracy in the United States.[2] Twenty-year-old Mark Zuckerberg launched Facebook from a dormitory room at Harvard in February 2004. Thirty-one-year-old Google marketing executive Wael Ghonim became the public face of a network of youthful organizers who took their Facebook organizing to the streets of Cairo on January 25, 2011, where hundreds of thousands of Egyptian citizens overthrew a thirty-year-old dictatorship in a little over two weeks.[3]

What is the common element in each of these stories? Knowing how to blog, tweet, wiki, search, innovate, program, and/or organize online can lead to political, cultural, and economic value. I emphasize the youth of most of these exemplars to highlight the unprecedented empowerment that digital know-how can grant—but participation in networked publishing is not limited to the young. According to Rainie,

The size of the mediasphere where people are telling stories, giving personal testimonies, contributing their ideas, and interacting with others has vastly expanded. Moreover, participation itself in the online world creates a distinct sense of belonging and empowerment in users. Pew Internet consistently finds that online participators—those who contribute their thoughts, rank and review material, tag content, upload pictures and videos—are at least a fifth of internet users on a range of subjects. For instance, 37% of internet users have made their own contributions to news coverage. Some 18% of online Americans have used social media tools to participate in politics. Some 20% of e-patients have contributed health related content. Finally 19% of internet users have posted civic and political material. They are the most active and engaged with their subjects and those are the most important precursors of personal influence.[4]

In the world of digitally networked publics, online participation—if you know how to do it—can translate into real power.

Participation, however, is a kind of power that only works if you share it with others. Even if their form of participation consists of ranting on their blogs, bloggers need publics to read, comment, and link to them. We participate online primarily for our own benefit, but if we use what digital culture leader Tim O'Reilly calls "architectures of participation,"[5] millions of individual acts of participation add up to a "participatory culture," as media theorist Henry Jenkins refers to it.[6] Communication media can make it possible for individual behaviors to add up to collective value by making it easy or affordable for people to do things together that used to be difficult or costly. Web culture has made it clear that if it is easy and inexpensive enough to contribute to cooperative enterprises, many people will choose to do so for a variety of reasons, including reputation, altruism, curiosity, learning, a sense of reciprocating value to a community that provides value, as part of a game, and contributing something for public use that you had to do for your own purposes anyway (open-source developers dub this "scratching an itch"). The Web, Wikipedia, open-source software, and even the notorious music file-sharing service Napster are all examples of this principle of "many people will cooperate if the medium makes it easy enough."

O'Reilly is a Web publisher, digital culture impresario, and one of the original organizers of the open-source movement. In his influential column, the Architecture of Participation, O'Reilly used the Napster case to illustrate how Web services can be designed to foster new forms of participation.[7] When a user of the now-defunct Napster service searched online for music to copy from another Napster user's computer somewhere (that's all Napster did: connect online music seekers with online music sharers), then the user who was searching automatically made their own music files available to other Napster users during that search and when downloading files. The Napster creators designed the service's architecture to work by designating the digital folder on a seeker's personal computer where the seeker stored downloaded music to be open to sharing—people who wanted to *seek* music via Napster were automatically recruited to *share* music via Napster.

Blogger and activist Cory Doctorow calls this phenomenon in which the act of using a resource supplies the very resource it uses "sheep that shit grass."[8] It does not cost me any extra effort to make public the bookmarks and tags I make for my own use in social bookmarking services—another example of an architecture of participation. Online discussions are by their nature architectures of participation. Web 2.0 enterprises build public and private value on platforms of architectural participation, the way digital technology builds on Moore's law of miniaturization of electronic devices. But architecture only becomes vital when humans use it to do things. An entire cultural ecology, teeming with subcultures, has grown up around digital participation. Understanding that cultural landscape—and why it is important to all humans, not just technology nerds—is the best way to start expanding your participation skills.

Although much of the practical knowledge I'm passing along to you comes from my own experience and research, in the matter of participatory culture I have been learning from Henry Jenkins for many years, ever since he helped me see how the amateur "fanzines" that fascinated me in the 1980s and 1990s foreshadowed the twenty-first-century blogosphere. In 2006, Jenkins along with Ravi Purushotma, Katie Clinton, Margaret Weigel, and Alice Robison coauthored a white paper titled "Confronting the Challenges of Participatory Culture: Media Education for the 21st Century."[9] Just as Engelbart's 1962 "Augmenting Human Intellect" electrified me when I discovered it in the 1980s, the work of Jenkins and his colleagues jolted me into radically reframing my approach to teaching and learning about social media production. Participatory culture as Jenkins and his team defined it is one with:

1. relatively low barriers to artistic expression and civic engagement,
2. strong support for creating and sharing creations with others,
3. some type of informal mentorship whereby what is known by the most experienced is passed along to novices,
4. members who believe that their contributions matter, and
5. members who feel some degree of social connection with one another (at the least, they care what other people think about what they have created).[10]

Digital participation literacy employs a toolbox of skills (persuasion, curation, discussion, and self-presentation foremost among them), and spans a range of involvement, from tagging a photo or bookmarking a site, to editing a Wikipedia page or publishing a blog. Like other social media literacies, there is a social element to participation literacy in addition to the individual how-to skills needed to participate. Crap detection is essential as well. Mindful participation also involves knowing how others profit from your unpaid labor, and making your own decisions about the value of what you get in return.

It doesn't take too many hours of Web surfing to realize that the technical ability to blog, upload a video, or edit Wikipedia doesn't guarantee that everyone will express themselves effectively. Most blogs are boring (which means that knowing how to find the ones that aren't boring is part of your infotention-filtering challenge). Too many comments in forums and blogs are not worth the time it takes to read them—or worse (part of attention self-training consists of cultivating the ability to stop reading something that promises to be toxic). Wikipedians are constantly healing the erasure or vandalism of their pages. And you could put yourself to sleep for the rest of your life—or run screaming from the room—by watching the truly terrible videos people put up for the world to see, or reading the excruciatingly mundane minutia of unfiltered Twitter.

The good news is that learning to participate effectively online (like learning attention and crap-detection skills) is a matter of mind-set and practice—and the payoff can be big. Knowledgeable online participation can help you land a job, find a mate, organize a movement, or sell a product or service. As citizens, professionals, and consumers, we hit it big, manage to get by, or fail utterly in large part because of our ability to connect and converse with others by way of digital networks, from LinkedIn to eHarmony. It isn't all about individual advantage. Done mindfully, digital participation helps build a more democratic, more diverse culture—a participatory one.

Participatory Culture

Full participation in contemporary culture requires not just consuming messages, but also creating and sharing them. To fulfill the promise of digital citizenship, Americans must acquire multimedia communication skills and know how to use these skills to engage in the civic life of their communities.

—Renee Hobbs, *Digital Media and Literacy*, 2010

Young and old around the world are re-creating the culture industry, or at least adding a new wrinkle to it, by having fun together. Don't let the unpolished nature of our amateur contributions block your view of the future. You should have seen what toys personal computers were in 1978 or what the Internet looked like in 1991. Viral videos and fan communities might not be the *New York Times* or CNN, but they signal an inflection point. Until it became possible to create videos with laptop computers and distribute them through inexpensive digital networks, video was something created by a small, well-paid guild of professionals and passively consumed by people who paid others for their prepackaged cultural products. Book, magazine, newspaper, radio, television, and music industries all conformed to the few-to-many model because of the high cost of owning technologies of production and distribution—printing presses, radio or television stations, recording studios, "ink by the carload," and fleets of trucks. Futurist Alvin Toffler predicted in his 1980 book *The Third Wave* that consumers were becoming "prosumers."[11] Thirty-five hours of video are uploaded to YouTube every minute.[12] Flickr reported its five-billionth photo uploaded in 2010.[13] The cultural fabric that emerges from the aggregation and interaction of millions of individually crafted productions in a many-to-many network of prosumers is bound to be different from the mass-manufactured fabric of a few-to-many culture.

Jenkins was an early proponent of the idea of a participatory culture. In an interview for this book, when I asked Jenkins what he meant by the term, he told me about a group of students that he and his colleagues had worked with:

We have been doing some projects around Wikipedia in schools, and one of the schools we worked with was using Wikipedia in the context of teaching *Moby Dick*. The students made changes to the official Wikipedia page for *Moby Dick*. Their changes were challenged by other editors, as often happens on Wikipedia, so the students marshaled arguments and evidence, and organized a defense of their changes, which eventually stuck. These students became part of the encyclopedia entry on *Moby Dick*. It was such an empowering experience for these kids to think that they could create knowledge that was accepted by what they saw as a fairly authoritative community and that became part of the public record of information about this particular book.[14]

Participatory culture is one in which a significant portion of the population, not just a small professional guild, can participate in the production of cultural materials ranging from encyclopedia entries to videos watched by millions. And it is a culture populated by people who believe they have some degree of power.

In 2009, Jenkins wrote on his blog,

This new emphasis on "participatory culture" represents a serious rethinking of the model of cultural resistance which dominated cultural studies in the 1980s and 1990s. Cultural resistance is based on the assumption that average citizens are largely locked outside of the process of cultural production and circulations; [Michel] De Certeau's "tactics" (especially as elaborated through the work of John Fiske) were "survival mechanisms" which allowed us to negotiate a space for our own pleasures and meanings in a world where we mostly consumed content produced by corporate media; "poachers" in my early formulations were "rogue readers" whose very act of reading violated many of the rules set in place to police and organize culture. [Jenkins was referring to fans who had reedited existing material such as *Star Trek* episodes to create new meanings; in the case of the "Kirk/Spock vidders," their retellings portray a love affair between Kirk and Spock—a rendition that wasn't exactly embraced by the series' original creators.] Increasingly, audience participation is factored into the business plans and are central to the design of media franchises; media companies alternatively seek to court and control an increasingly unruly audience as fans and other consumers recognize that collectively we exert much greater influence on the cultural agenda and are helping to generate the content that others are consuming. As consumers and citizens have taken media into their own hands, they are becoming more aware of the economic and legal mechanisms which might blunt their cultural influence and are defining strategies for using these new platforms in ways that promote their own interests rather than necessarily those of their corporate owners. In this new context, participation is not the same thing as resistance nor is it simply an alternative form of co-optation; rather, struggles occur in, around, and through participation which have no predetermined outcomes. Both producers and consumers may now be understood as "participants" in this new media ecology, while recognizing that they do so from positions of unequal power, resources, skills, access, and time.[15]

Social media literacies are potential tools and weapons in these struggles over participation.

When I visited Japan in 2001 to research texting culture, I found my way to Ito, the Stanford-trained anthropologist at Keio University who was studying the way teenage girls in Tokyo used their mobile devices. She's become a trusted source for firsthand reports from the leading edge of participatory culture. What she is observing now in the United States might portend a new mode of learning in which education more closely resembles a many-to-many network of colearning enthusiasts than a roomful of students and a one-to-many teacher. Considering the rates of change in the knowledge-based professions, learning networks won't be limited to schools populated by young people but instead will be essential survival tools—and perhaps sources of enjoyment rather than onerous chores—for those want to keep up with what's happening at the edge. For three years, Ito led a team at the University of California in the most comprehensive investigation ever conducted into the way young people use new media, culminating in a 2009 report, *Living and Learning with New Media: Summary of Findings from the Digital Youth Project.*[16]

One of the most important understandings learned from Ito's research was that young people were using digital participation tools in ways that are not only social but also deeply involved in learning and creating:

For the past few years I have been looking for learning in somewhat unexpected places—in young people's social and recreational practices surrounding new media. I have been guided by the belief that interactive, digital, and networked forms of media are supporting new forms of engagement with knowledge and culture with unique learning dynamics. My fieldwork is indicating that a key trigger for these learning dynamics is the peer-to-peer traffic in media and knowledge that accompanies young people's engagement with culture and knowledge that they are passionate about. As they become more pervasive in our everyday lives, networked and digital media become a vehicle and an infrastructure for this peer based learning and sharing. First, there is very little explicit instruction, and learning happens through process of peer-based knowledge sharing. People engaged in a practice seek out information or knowledgeable peers when it becomes relevant to their work, and in turn, they help others when asked. Although there are people acknowledged as experts, they are not framed as instructors. . . . Finally, these environments are based on ongoing feedback and reviews of performance and work that are embedded in the practices of creation and play. These groups also have contexts for the public display and circulation of work that enables review and critique by their audiences. Competition and assessment happens within this ecology of media production and consumption, not by an external mechanism or set of standards. In other words, individual accomplishment is recognized and celebrated among peers in the production community and other interested fans, providing powerful motivation for ongoing learning and achievement.[17]

Ito studied online fan cultures such as the young people who create online forums devoted to the face-to-face card game Yu-Gi-Oh. She also observed the community that gathered around the making of anime music videos (repurposing Japanese anime animations by reediting them to new sound tracks), and vidders who reedited mainstream cultural products to create entirely new meanings and grow a decades-old worldwide community. Ito notes, "Participatory media cultures like anime fandom have unique dynamics that are based on a very particular genre of participation embedded in a whole fabric of community life and social communication. While the specificities of these practices are unique to this fandom, the turn towards amateur cultural modes aided by digital media and networking is a much broader sociotechnical trend."[18]

Ito's team discovered some connecting patterns in their observations—"genres of participation"—that can be useful to digital participants of all ages today. They found that digital participation differs in crucial ways, depending on whether it is "friendship-driven" or "interest-driven" participation. "The dominant mode of friendship-driven participation is what kids call hanging out. This is the relatively unstructured, often impromptu ambient social activity where so much of youth socialization happens."[19] Friendship communities are densely connected—most know each other in other contexts—and highly local. Facebook is a friendship-driven digital place to hang out for many young people. As part of their online socializing, youths learn from and teach each other methods to create as well as share media and lore. Knowing where to find the next cool music file to download is a form of social currency, as is being able to reedit it. Pictures, wall comments, status updates, and videos are all part of hanging out digitally, all deeply intertwined with face-to-face social life. Fans fall on the interest-driven side and not the friendship-driven one, and the interest-driven kids are a minority, while friendship-driven practices are majority mainstream practices.

Interest-driven communities are not formed from people who already know each other, and use digital media to hang out and share media online; rather, they are created by people who had not previously known each other but use digital media to find each other, hang out, and share the products of their mutual interest. These communities have a mix of old-timers who have come to know each other and newbies who are learning their way around. The exercise and exchange of digital media skills is what Ito and her team label "messing around" through the informal, playlike exchange of cultural products, such as fan videos and information such as Pokemon arcana. These creative genres of messing around may foreshadow the way many people will learn necessary skills in the future. All the cool kids and wannabes are likely to be found in a friendship-driven community. Interest-driven participation, however, is full of people who are not popular or mainstream in their local youth culture,

the kids we see at the margins of teen social worlds. This is about kids with passionate interests and serious hobbies finding peers online. It is not about the given social relations that structure kids school lives, but it's about expanding an individual's social circle based on interests. Kids who have a strong interest-based orientation will often talk about how they don't like to participate in sites like MySpace and prefer online forums that are focused on interests. These are the kids who are creating

YouTube videos, leading guilds in online role playing games, remixing movies and videos and sharing them online, or participating in Harry Potter fan fiction groups.[20]

The activities that interest-driven groups engage in involve more serious learning and deeper involvement in the crafts of their subculture, generating more sophisticated and specialized roles, methods, products, and tools—what Ito refers to as the genre of "geeking out."[21] Thirty years ago, two of the people who created today's digital tools, Gates and Jobs, were exactly the kind of interest-based subculture fans Ito portrays, except their fringe subculture consisted of teenage enthusiasts who liked to geek out by building their own computers. Whether they are driven by friendship or interest, the young people across the United States who Ito's team studied, representing a broad demographic sample of the population, use media sharing and production as a form of social currency. The communities they develop are based on creation, exchange, collaboration, and critique of media created by participants. Ten years from now, when these kids enter the workforce and political arena, we'll find out whether people who grow up in a participatory subculture change their ways or the dominant culture. The skills they are learning provide a model for the spectrum of skills that all mindful digital participants presently can deploy for their own benefit and the public good.

Participation Skills: Skating up the Power Law of Participation

From clicking on a link or "like," or "favoriting" a tweet or video, to organizing collective action in an election or revolution, lightweight forms of participation can lead to more engaged and creative involvement—and individuals can take small steps that become part of big things. I maintain several blogs myself, tweet regularly, organize online learning communities, bookmark socially, upload and tag my photos and videos, comment on other people's blogs, curate topics such as "infotention" through Scoop. it, and share video and slide show links. Am I drunk on participation, or cashing in on it?

Fred Turner, Stanford professor as well as head of Stanford's program on science, technology, and society, claims that people like me both contribute to the commons and profit from it by being "network entrepreneurs."[22] By this, Turner means that we network entrepreneurs benefit in reputation and audience/public by giving our products freely to our networks, and that when we act to connect previously disparate networks, we put ourselves in a position to profit from the connection. I don't think self-interest is a bad thing, especially when it drives the creation of public goods as a

by-product. I believe we'll all be better off when more people learn to be good network entrepreneurs; you can't succeed at network entrepreneurship unless you consistently enrich your network. (I'll offer more about network entrepreneurship when I look at network literacies.)

Much can be learned about how to engage successfully in both lighter weight and more heavily committed forms of participation by exploring the blossoming of the blogosphere. In the mid-1990s, I was part of a small subculture of online self-publishers who wrote something new on our home pages, as Web sites were called at first, practically every day. I had to wrap my writing in hypertext markup language and upload the file with a special command known as file transfer protocol.[23] I then had to issue another command through the telnet protocol so the rest of the world could see the file. In 1999, a free, Web-based application called "Blogger" was invented—originally as a way for people in a small Web company to keep track of each other's work. It took one minute to create a Blogger account, at which point all that tens of millions of people and I had to do was to type our words into a form on a Web page and click the "publish" link. The culture of self-publishers that flourished in the early 2000s came to be known as the blogosphere. I won't concentrate here on the impact of blogging but instead will refer my readers to Scott Rosenberg's excellent book, *Say Everything: How Blogging Began, What It's Becoming, and Why It Matters.*[24]

The connection between blogging and participatory culture grows from the way the experience of blogging changes the blogger. Rosenberg explains the importance of the experience of participation on his blog, *Wordyard.com*:

Learning to make things changes how we understand and consume those things. . . . Writing for an audience is a special and important sub-case: it's writing with feedback and consequences. Doing it yourself changes how you think about it and how you evaluate others' efforts. The now-unfashionable word "empowerment" describes a part of that change: writing is a way of discovering one's voice and feeling its strength. But writing in public involves discovering the boundaries and limits of that power, too. We learn all the different ways in which we are not the center of the universe. So when I hear the still-commonplace dismissal of blogging as a trivial pastime or an amateurish hobby, I think, hold on a second. Writing—making texts—changes how we read and think. Every blogger (at least every blogger that wasn't already a writer) is someone who has learned to read the world differently. The person who has struggled to turn a thought into a blog post, and then seen how that post has been reflected back by readers and other bloggers, is someone who can think more creatively about how sharing might work at other scales and in other contexts. A mind that has changed is more likely to imagine a world that can change.[25]

Blogging is a way to find your voice and public, connect with like-minded communities, improve your digital profile, influence others, and contribute to the commons. For practical advice about blogging as participation, I begin with the notion of *voice* and four of the genres of blog rhetoric I have introduced to my university students: blogger as *filter, connector, critic,* and *advocate.* Those are not the only roles bloggers take on, but they are fundamental and not hard to learn. That some people blog better (and more popularly) than others should not stop anyone from communicating with an authentic voice to their public, just as the awesome talent of Michael Jordan should not prevent others from playing basketball.

• We can all filter (more on that when I talk about curation) by simply sharing what we find when we pursue our interests, selecting the best stuff for our own edification, and then recommending it to others
• Making connections is a learnable skill that is amply rewarded by networked publics—and every blog post that includes a link makes a connection
• "Everybody is a critic" is a cliché, and again, although some will be better critics than others, the aggregated critiques of ordinary people directed toward politicians, products, or the service at the restaurant they ate at last night can add up to a valuable public good
• Whether it is opposing or proposing leash laws for dog owners (a hot topic in my geographic community), supporting or defeating a political candidate, or raising money for medical research, everybody at some time or another has a cause to advocate, whether or not they have a platform for broadcasting their views

I came to understand the importance of developing a public voice online through an essay by Phil Agre, whose email list about digital culture predated the term digital culture by at least a decade. Although it's ancient by Web reckoning, I still assign my students Agre's 1999 "Find Your Voice: Writing for a Webzine."[26] Agre asked that his draft essay not be quoted, so I will paraphrase it. A private voice is about self-expression and does not seek to engage readers in discussion of broader issues. A commercial voice is about producing an effect, whether it is selling a product or entertaining an audience. A public voice is somewhere between the two, and strives to remain true to the communicator's own experience while joining with others in a public conversation—what Rosenberg calls writing in public about issues that concern publics. Voice is the foundation of presenting an identity online—what sociologist Erving Goffman dubs "the presentation of self."[27] Your voice is what distinguishes your views from others who hold the same opinions. Your voice connects who you are with what you care about—and the other people who care about the same things.

Many bloggers serve as "intelligent filters" for their publics by selecting, contextualizing, and presenting links of particular interest for that public. In this regard, a "public" differs from an "audience" because you, in your role as a blogger, keep a community of peers in mind when you write—people you may or may not know personally, but who not only read but also could potentially respond to what you write, who might act on your advice, and who might join you in discussion and collective action. The public you choose to address could be one in the sense of a political public sphere that undergirds democracy—the communications you engage in with your fellow citizens about issues large and small, with whom you share responsibility for self-governance. The public doesn't have to be political, however. It could be an engaged community of interest—others who share your profession, avocation, or obsession. When fans begin writing fan fiction, or remixing and sharing cultural content, they act as culture-producing publics. AIDS patients organized collective action that influenced research funding and the pharmaceutical industry, thereby creating an effective public through their discussions about their mutual interest.

To start blogging as a filter blogger, first define in your own terms the public you want to engage. Keeping that public in mind, post a link in a blog post that connects to any site on the Web—a blog post, a mainstream news item, a Wikipedia entry, an online community or marketplace, or audio or video content—that has the potential to enhance that public's knowledge, incite that public to take action, or provoke that public to respond to you. Don't stop with sharing the link. Add context. Why, as an expert on wedding cakes or tropical plants, do you find the linked site to be useful or interesting to other wedding cake or tropical plant mavens? All of us belong to multiple communities of interest, each of us is especially interested in something, and many of us pursue our interests diligently enough to lay claim to some degree of expertise. Don't carry on about how expert you are; demonstrate your expertise in a useful way by explaining why it is worth your public's time and attention to click on the link you offer. Do you regularly follow other bloggers who care about the same issues as you? Add them to your "blogroll" (public list on your blog of other blogs you follow and recommend) to aid others who are seeking networks of expertise. When you gain the confidence to engage your public, open your blog for comments. Learn to ignore the trolls (people who make outrageous or nasty comments in order to provoke a reaction). Pay attention to the critics; they are your teachers, giving you free advice. And cultivate the fans who think enough of you to make suggestions for you to share with your public. Don't forget that others will read your responses to comments when they try to evaluate your credibility.

When I started using blogs and wikis in classrooms, I discovered Will Richardson's book, *Blogs, Wikis, Podcasts, and Other Powerful Web Tools for Classrooms*, and started following his blog and Twitter stream—adding him to what I didn't yet call my PLN.[28] Richardson introduced me to the blogging rhetoric he labels "connective writing." As Richardson describes it in his blog, connective writing is particularly suited for this form of social media:

A new type of writing that blogs allow, one that forces those who do it to read carefully and critically, one that demands clarity and cogency in its construction, one that is done for [a] wide audience, and one that links to the sources of the ideas expressed. . . . I'm talking about something uniquely suited to blogs. I'm talking about this post, about our ability to connect ideas in ways that we could not do with paper, to distribute them in ways we could not do with the restrictiveness of html, and to engage in conversations and community in ways we could not do with newsgroups or other online communities before.[29]

Think of connective blogging as taking filtering one step further. Go beyond telling your public why it should click on a link. Reflect on the resource or idea you are linking to. Read carefully and critically, and write something about what the linked source means. Push it further by thinking of a broader context and bringing in another link. Make the connection between these links clear to your public. Again, this is a way that the experience of blogging about something can change (deepen, broaden, and challenge) the way the blogger thinks about that thing. Richardson quotes Ken Smith, a writing instructor at Indiana University: "Blogging, at base, is writing down what you think when you read others. If you keep at it, others will eventually write down what they think when they read you, and you'll enter a new realm of blogging, a new realm of human connection."[30]

Even a casual examination of bloggers' critiques of books or movies, politicians or legislation, and restaurants, hospitals, or automobile repair shops makes it clear that many blog critics don't know what they are talking about. That's not the point. Rather, if you know how to find the many people who *do* know what they are talking about when you seek a restaurant, surgeon, or beach reading, you can use the collective intelligence of the critical blogosphere as your personal guide. The presence of bad information in the critique pool won't destroy the aggregate value if human (curated) or automated (algorithmic) methods emerge to enable the good information to rise to your attention. If you can find a couple of bloggers whose opinion you've come to trust about a subject, you can build an expert network by following the others in their blogrolls. You don't have to hate a meal or find an incorrect statement of fact to be a critic. Debate the logic or possible bias of an author. Make a counterargument. Indicate what the author leaves out. Voice your own opinion in response.

What information would you link to support your claims? When you compose and publish your critique, you've not only contributed to the collective intelligence about your field, you've honed your ability to think critically and present your criticism persuasively. The ability to analyze, investigate, and argue about what we read, see, and hear is an essential skill. Some bloggers do spread the most outrageously inaccurate and fallaciously reasoned information; it is up to the readers—and most significantly, other bloggers—to actively question the questionable publicly, where others can benefit.

The blogger as advocate, like the other rhetorics of blogging, can change both the blogger and the blogger's public. During the 2004 elections, Sinclair Broadcast Group announced plans to air what Democrats considered to be a blatantly partisan documentary about John Kerry, one of the candidates for U.S. president. By forcing sixty-two stations to run the documentary in the days before the election, Sinclair had the potential power to sway a tight election. Left-wing bloggers crowdsourced the task of finding out how to contact all of Sinclair's local sponsors by dividing up the research and parceling it out among hundreds of bloggers, and then organized a boycott that dropped Sinclair's stock price by ten points, forcing the media conglomerate to back down on its plans.[31] Similarly, right-wing bloggers crowdsourced the task of debunking claims by *CBS News* about candidate George W. Bush's Vietnam War record and ultimately cost Dan Rather his job.[32] Whether or not a particular critical blog post initiates the kind of political effects that the Sinclair or CBS bloggers did, every blog advocate understands how it feels to not just live with the consequences of democratic policymaking but instead to take an active stance. Whether or not most bloggers have any real influence, the blogosphere has proved to be influential politically, with an attention-seeking ecosystem that enables the discoveries of previously obscure bloggers to be linked, quoted, and reblogged by one of the stars whose blogs are read by millions.

If you aren't ready to blog or that form of publishing doesn't inspire you, you can participate in ways that don't involve as much public disclosure. The "power law of participation" (an idea originally published by blogger and social media entrepreneur Ross Mayfield) illustrates the ways you can participate by reading, tagging, favoriting, subscribing, commenting, or sharing.[33] Commenting, on the low-commitment part of the curve, is an easy way to start interacting with people whose publications you admire. The blogosphere is full of ignorant, less-than-useful, even hateful remarks.

Comments that actually engage the ideas of a blog, argue in a civil manner, offer counterevidence to claims, encourage the blogger, or provide additional useful information are golden to bloggers as well as those of us who read comments as part of both information gathering and crap detection. Communities of bloggers and commenters have created valuable knowledge repositories and social capital; again, finding the good ones is an infotention skill. Commenting is also a way of associating your name with findable content that improves the quality of your reputation plus makes you visible to others who share your views or interests. Uncloaking yourself digitally by posting a comment can have a profound effect on how you perceive yourself. When an admired blogger responds to your remark or other commenters engage you, you are no longer just a reader; you are a member of a network—perhaps, eventually, a member of a community.

You can participate in ways that fulfill you and contribute to others by simply sharing the best media you find. Sharing—sending a link or embedding a video—is not only a way of enriching your friends. It's also a contribution to a commons. When you are trying to decide which one of a number of videos are of higher quality or are more useful than other similar ones, look at how much a video is shared—often a better measure of its relevance than the number of views. When you share, you make personal connections with people you know while enhancing the experience of strangers who follow your blog and contribute to the collective evaluation of online media. You exercise mindfulness when you ask yourself whether you are enriching someone or stealing part of their attention when you share a video of a revolution or a cute kitten. You exercise crap detection when you ask yourself whether the story or media object you are sharing is legitimate before you spread it around. When you share, favorite, like, bookmark, or tag something you find and evaluate online, you're moving into curation, another participation skill that enriches both the curator and commons.

Curation Is Short for "We're All Each Other's Filter"

A tweet is an atom. A photo on Flickr is an atom. A conversation item on Google Buzz is an atom. A Facebook status message is an atom. A YouTube video is an atom. Thousands of these atoms flow across our screens in tools like Seesmic, Google Reader, Tweetdeck, Tweetie, Simply Tweet, Twitroid, etc. A curator is an information chemist. He or she mixes atoms together in a way to build an info-molecule. Then adds value to that molecule.

—Robert Scoble, "The Seven Needs of Real-time Curators," 2010

People have always made choices about what to pay attention to. In the online world, they also make choices that influence what *others* pay attention to. That's what people mean by the word curation in reference to online behavior. The curator role used to be reserved for the people who ran museums, but the term has been revived and expanded to describe the way populations of Web participants can act as information finders and evaluators for each other, creating through their choices collections of links that others can use.

The voluntary curation contribution of every person who ever puts a link on a Web site, blog, or tweet is what enables Google to not only find Web sites that mention specified strings of words but also rank the sites in order of popularity. Popularity turns out to be a pretty good, though far from infallible gauge of a site's potential value. Crap-detecting skills are still necessary to separate the accurate from the merely popular. Google itself is not the curator; we are. Every time a person references a link, they help to curate the Web. In 2010, Facebook recognized the power of curation by making available to millions of sites a "like" button that Facebook users could click on to recommend that site to others. Best-selling author and Web entrepreneur Seth Godin foresees curation as an emerging position of power: "If we live in a world where information drives what we do, the information we get becomes the most important thing. The person who chooses that information has power."[34]

Curation is a form of participation that is open to anyone who might not want to blog, tweet, or update a Facebook profile but instead are happy to bookmark, tag, or like other people's digital creations. Judgment, taste, depth, and breadth of knowledge can be an asset, a public good, and a commodity. People can gain attention, admiration, collaboration partners, professional reputations, and business relationships by becoming known curators. Curation has been an interest of mine since I wrote in 1987:

In my virtual community, we don't have software agents (because they don't exist yet), but we do have informal social contracts that allow us to act as software agents for one another. If, in my wanderings through information space, I come across items that don't interest me but which I know one of my group of online friends appreciate, I send the appropriate friend a pointer to the key datum or discussion. This social contract requires one to give something, and enables one to receive something. I have to keep my friends in mind and send them pointers instead of throwing my informational discards into the virtual scrap-heap. It doesn't take a great deal of energy to do that, since I have to sift that information anyway in order to find the knowledge I seek for my own purposes. And with twenty other people who have an eye out for my interests while they explore sectors of the information space that I normally wouldn't frequent, I find that the help I receive far outweighs the energy I expend helping others: A perfect fit of altruism and self interest.[35]

In pursuit of latter-day master curators, I interviewed one of the first journalists whose curation brought him celebrity among the digerati, Scoble.[36] I had the privilege of watching him in action during a two-week "traveling geeks" tour of the United Kingdom. Like all great curators, Scoble is first (and energetically) an enthusiast and friend of enthusiasts. He loves to find intriguing people, ask them questions, and tell their stories. Scoble is a fan of social media, Web entrepreneurs, and the latest digital tools. He blogs, microblogs, favorites tweets, and likes Web sites, and tags and bookmarks thousands of people, enterprises, media, and Web sites about those subjects. And because he understands the literacies of attention, participation, crap detection, collective intelligence, and social network dynamics, Scoble can do magic in public—like the time he beat the U.S. Geological Survey (USGS) to the news of an earthquake in China. According to Scoble, his passion for curation started in college. "In 1991–92," Scoble recalls,

I ran the Associated Press wire machine at San Jose State University, and fell in love with that. I was running the machine when O. J. Simpson was found not guilty. There was something like 650 stories from hundreds of journalists who had been at the courthouse. The *San Jose Mercury News* picked two or three stories to put in the newspaper, but I always wanted to have more. Today, Twitter replaces the AP wire machine for me. I have a Twitter list of all the world news brands, and now I can see world news happen in real time. Most people can't put up with all that information streaming at them, so I try to pick interesting things out of the flow for a geek audience; I am curating for geeks or people interested in technology. It's like a newspaper editors' job. I enjoy picking out interesting things and seeing patterns that other people don't see in the news.[37]

Scoble came to beat USGS to the Chinese earthquake because, as he told me,

I watch tens of thousands of people on Twitter. I was just watching as entertainment one evening and I saw five different people say something about an earthquake— and those people were located in five separate cities in China many miles apart. So I kept refreshing the USGS Web site and Twitter search. I always start out skeptical because it's not fun for your brand to be pushing along stuff that doesn't prove to be true. Once the USGS confirmed the quake, the people who had reported it on my Twitter feed became authoritative on that subject. I immediately started watching everything they wrote. One of them started linking me to pictures, videos, and news reports—forty-five minutes before the Associated Press and CNN admitted there had been an earthquake in China.[38]

Scoble's advice to novice curators is to

start out by building a Twitter list of people you trust and you like reading. Don't add any person or organization to the list until you understand who they are and you've watched them for a little while. I look at these lists as funnels. Who do I put in the funnel? I want people I trust for some reason. On my main Twitter list I follow all geeks. On my tech-influencers list, you actually have to have some influence in the technology world to get on it. Try to meet and interview the people on your list who prove to be the most interesting and useful. Don't try to compete with me or the *Huffington Post*. Find a small niche. Really understand it. If you wanted to study the White House, study what a pet is doing at the White House. Pick something very specific and become the world authority on what the pets are doing at the White House. Totally cover it. Own that small niche.[39]

Scoble codified the emerging craft of curation on his famous blog, Scobleizer.com. According to Scoble, the "Seven Needs of Real-time Curators" include (in my words):[40]

1. The need to *bundle*. Scoble bundled together tweets about the Chinese earthquake, along with links to videos and pictures. Tools like Storify.com, curated.by, and Scoop.it are emerging to make bundling and other aspects of curation as easy as blogging.
2. The need to *reorder*. The order in which you found information, or the order in which it flowed to you, is not often the order you want to present to your public. Arranging the most important items in a useful order is part of the value curators add.
3. The need to *distribute* bundles. Curation tools increasingly include automatic blogging, so each bundle can have a URL. Without a URL, there's nothing to link to, and links are the Web's connective tissue and social currency.
4. The need to *editorialize*. "So, now we have a bundle of Tweets, YouTube videos, Flickr photos, Google Buzz items, [and] Facebook status messages. We've seen a new pattern in the world and now we want to explain our view of that pattern."[41] I don't just want Scoble to bring a new app to my attention. I want to know what he thinks of it, and why.
5. The need to *update*. Stories and bodies of knowledge change over time—some more quickly than others. In an era of information flows and activity streams, curated bundles have to be updated (including as part of the bundle RSS feeds, since specific tags are one way).
6. The need to *invite participation*. Enable reader comments on your blog (this can make you vulnerable to spammers who auto comment their uninvited ads; Scoble leaves comments for any story open for only thirty days for this reason, and other means can be used to filter out spammers).

7. The need to *track your public*. TweetMeme.com can tell you how many times your tweet or bundle has been retweeted (copied and published by others via Twitter), and Google Analytics can provide an entire dashboard. URL-shortening services like goo.gl and bitly enable you to keep track of how many times your shortened link is clicked on.

I was fortunate to also have met the other person I consider to be a world authority on curation, Luigi Canali de Rossi, known to the world as "Robin Good." I'll never forget seeing Rome from the back of his motorbike. And the RSS feed from his MasterNewMedia superblog is in the top-left corner of my dashboard—the place I position the most useful, timely information flows.[42] Good has published a detailed series on what he calls "newsmastering" or, like Scoble, "real-time news curation." I'll distill and quote from Good here, but I recommend his comprehensive "Real-time News Curation, Newsmastering, and Newsradars: The Complete Guide, Parts 1–6."[43] Good listed the following attributes of a good curator (again, paraphrased— and note the overlaps with Scoble's advice about curation):

1. *Subject matter expertise.*   This involves not just cold knowledge but also a passion for the topic.
2. *Relevance.*   "The result of subject matter expertise coupled with a very good understanding of the audience one is trying to serve."
3. *Trust.*   "Repeated relevance," Good succinctly states, leads to trust. It's why you want to double check before passing along bad info and damaging the trust your public has granted you.[44]

Next, Good listed the steps in a curatorial work-flow process. If you want to go beyond basic literacy to become what Good calls a newsmaster, here is your guide:

1. *Identify niche.* As Scoble also pointed out, focusing on a specialty is the way to distinguish yourself.
2. *Select.* Make sure to identify your sources. The kind of exploration in search of expertise I'll discuss in regard to PLNs will be relevant to this step.
3. *Search framework.* Set up a dashboard, and use Twitter and social media searches ("radars") to monitor information flows about your niche.
4. *Reach out.* Network with "reporters, journalists, passionate users, influencers and experts in your topic niche."[45]
5. *Aggregate.* Create a single outgoing feed about your niche, compounded from your selected, filtered incoming feeds
6. *Filter.* Use PostRank and other filters to weed out the least useful input before it reaches you. Protect your public from spam by using the spam-filtering tools that most comment systems now offer.

7. *Select* stories. Here's where you apply your taste and expertise.

8. *Verify*. Crap detect before you send information out to your public—lest they crap detect *you*.

9. *Edit*. Introduce, summarize, copyedit, and check your references and citations.

10. *Contextualize*. Learn the art of the snippet that enables your public to zero in on the reason why you've selected this item for their attention.

11. *Spin*. Add perspective, state your opinion, and explain the big picture.

12. *Title*. In the world of search engines and human infotention, you need to concentrate on the findability and spreadability of the titles you give every outgoing story or bundle.

13. *Credit*. Don't just link; reference. Provide citation information and credit your sources.

14. *Sequence*. "Time is not always the best sequencing variable."[46]

15. *Organize*. Tag intelligently.

16. *Update*. Turn inert content into an info flow by adding an RSS feed to your bundle that updates with links you bookmark under specific tags.

17. *Disclose*. "Be upfront about your focus, mission, and personal profile information. Disclose as much of this info as possible letting your readers-subscribers know what is your topic, perspective, or editorial take on it and your specific background and expertise. Make commercial partnerships and sponsors you have publicly known."[47]

18. *Syndicate*. Let your public know how to subscribe to your updates through RSS.

19. *Feedback replay*. Respond to suggestions, critical feedback, debates, and invitations from your public.

20. *Monitor*. All kinds of analytic tools enable you to find out more about which of your stories are most clicked, favorited, liked, or shared.

21. *Refine and improve*. If you are serious about curation, this cycle makes it possible to put attention into continuous improvement, with the help of your public.

If you want to curate, keep Scoble's and Good's advice in the forefront of your attention, and apply the practices they recommend. Two ways to get started are by learning to tag intelligently (and follow tags to find the information you seek) through social bookmarking and making public lists, such as Twitter lists.

Search engines have replaced much of our previous practice of indexing information within folders, often nested within other folders. If you can remember the exact name of a document or a unique combination of words you know are on a Web page, there's no reason to store a reference. But that doesn't help to store for your own reference a collection of previously vetted Web sites on a particular subject or curate a library of resources on a subject for others. A programmer who worked on databases for a financial company, Joshua Schachter, cooked up a quick program that allowed him to store URLs for Web sites along with his summary or a snippet of text from each site, attach multiple keywords to each bookmark, and then retrieve the links later by looking up the collection associated with any of the tags he used. A link to a video by Marshall McLuhan about media could be stored and then retrieved via either "McLuhan," "video," or "media" tags. Instead of choosing which folder to put a bookmark in, tagging makes it possible to put it in as many categories as I'd like. When his friends started asking for copies of his program, it was easy to make it possible for any of his friends to make his bookmark choices, summaries, and tags available to others. Schachter's personal knowledge-management utility grew into a Web service used by millions of people, now known as delicious.com, and the genre he created is "social bookmarking," a practice that allows those with sufficient attention, collaboration, and network know-how to multiply the value of a public good while serving one's own self-interest.

Social bookmarking makes possible discovery as well as curation. I can use my own particular expertise to cull out the best links on a topic I know about, and enable others to access my vetted collection;, I also can browse through other people's collections of resources to find good information about particular topics. Give me a subject that can be described in a word or two, like "media literacy" or "molecular gastronomy," and I can pinpoint within minutes a crucial online resource or two, and begin to assemble a personal network of relatively trustworthy expert sources on the topic. In that way, social bookmarking is a vast knowledge commons, free for the harvesting. And as I accumulate my own collection of links to, for example, material about wikis, I can make a resource like http://www.delicious.com/ hrheingold/wiki available to anyone who inspects the bookmarks tagged "wiki," those who browse my own publicly viewable bookmark collection, and those who subscribe to my wiki tag and receive an RSS update whenever I tag a new bookmark with "wiki." And when I want to point my students or a colleague at my bookmarks for using wikis in collaboration, I can send them to http://www.delicious.com/hrheingold/wiki+collaboration.

The World Wide Web has become, to those who know these skills, a personal learning tool and instant expertise finder, and at the same time, a free, global, automatic aggregator of facts, documents, and media that grows in value as well as volume every time every social bookmarker tags a site for later retrieval. And when I want to find good resources on neuroplasticity, I look for the collection of sites tagged with that keyword and inspect those that have been bookmarked by people whose expertise I trust—or those that have been bookmarked by the largest number of strangers. When I'm in learning mode, I look to see who else bookmarked pages I find valuable and who else collects information about a specific tag. I look to see what else they bookmarked, who is in their network, and whether I want to add their sources to my network of inputs. Social bookmarking service Diigo. com enables users to highlight passages and send to their networks links to highlighted versions of Web pages; its users can form groups and leave "sticky notes" for other group members (and reply to them) on Web pages. In effect, people can write in the margins of Web pages for only those other people who want to see their highlights and notes.

Tagging isn't just a way to participate. It's the fundamental building block of a whole new way of aggregating and organizing knowledge. David Weinberger's book (recommended for those interested in the ways the Web has changed knowledge) *Everything Is Miscellaneous* proposes that people organized the world in categories and subcategories (what is known as an ontology) for thousands of years because they didn't have search engines.[48] Only a few things in the world neatly match a nested or branching ontology, Weinberger argues (convincingly). Everything else is messy. A bear is a mammal, a toy, or a kind of market. Tag and search is more natural, more suited to the world, than categorize and pigeonhole.

When millions of people tag, categories emerge, and entities can easily be stored and found via multiple categories—an organizational form that has come to be known as a "folksonomy." By their nature, folksonomies have some advantages. Ontologies are created by elites who sometimes don't know what is happening in an adjacent discipline; folksonomies are democratic and broader, which means that noise enters the system, and also that the categorization will always be more inclusive (there are ninety categories for Christianity in the Dewey decimal system, and ten for all other religions combined). Entire populations will always be able to keep up with changing topics faster than panels of experts. Yahoo! learned that when it tried to index the Web by paying geeks in cubicles to classify sites by hand. Just as I find experts by searching the networks of people who bookmark tags of interest, folksonomies allow for discovery as well as retrieval—finding relevant knowledge (and knowledgeable people). Similarly, folksonomies enable networks and communities of interest to form among people who had not previously known one another.

Twitter lists are a lightweight way to practice curation, as Scoble advises, with a rich variety of niches to explore. Tell the world that the people on your list are particularly worthy of attention on a specific topic—directory Listorious.com has twenty-four separate categories of lists, some of them broad enough (sports or technology, for example) to encompass numerous subcategories. Publish your lists and submit them to list aggregators. When another Twitter user wants to follow only the most knowledgeable beekeepers or badminton journalists globally, and turns to your list, your judgments provide a filtering service for others—and if you do it well, word of your expertise will make its way around the network, since such news is the currency of online social capital.

The magic ingredient in tagging and folksonomies is the concept of metadata, which means information about information. It turns out that on the wild Web, information isn't valuable even if it's relevant if you don't know how to find it. Tags and other metadata make it easier to find relevant information. The individual worth of Flickr is that I can post pictures of my puppies. The metadata that millions of people have contributed to billions of images—making it possible to form communities around sunsets or custom automobiles, or for volunteers to classify the U.S. Library of Congress collection of images—is the added value that the owners of Flickr (currently Yahoo!) or YouTube (owned by Google) harvest as financial profits.

I personally think I get a pretty good deal. I get a free or inexpensive service; we all get an image commons, bookmark commons, video commons, and slide commons. But the trade-offs I make are perhaps more beneficial to me than those made by a less network-entrepreneurial person than myself. And the public resource is no longer a commons when it is subject to the decisions of the platform owners. For instance, Yahoo! recently caused a stir when it leaked its intention to divest itself of the Delicious.com social bookmarking service. Big corporations whose stock we don't necessarily own profit from our labor. But is it labor? Or is it play? The blurring of that distinction has led some people to refer to this behavior as "playbor."

Playbor: Do You Know Who Profits from Your Participation?

"In an information-rich world, the wealth of information means a dearth of something else: a scarcity of whatever it is that information consumes. What information consumes is rather obvious: it consumes the attention of its recipients. Hence a wealth of information creates a poverty of attention and a need to allocate that attention efficiently among the overabundance of information sources that might consume it.

—Herbert Simon, "Designing Organizations for an Information-Rich World," 1971

Attention is not only an inward-pointing instrument that you can learn to control; it is an economic factor that others seek to control. Every Facebook update, tweet, Flickr photo, and YouTube video you upload contributes clues to what kind of media and media content might get your attention—clues that are detected as well as analyzed by enterprises, individuals, and political actors who want to sell you a product, service, or idea. For example, car rental companies know that "people who have recently read online obituaries tend to be higher purchasers of weekend rental cars."[49] For every word you can define correctly at Freerice.com, the site donates ten grains of rice to the UN World Food Programme. The site is sponsored by ads that pay off only when people click through and make a purchase. This action is tracked by putting a cookie file on the player's computer—potentially a source of other valuable information for advertisers.[50]

The aggregated by-products of digital participation add up to a marketable commodity, and participation also adds value in more diffuse ways such as sustaining entertaining conversations in online communities (the customers create the value that attracts more customers), or contributing signals to info finders and crap detectors (like PageRank). Although I am convinced that participation is empowering to knowledgeable participants in digital cultural production, I believe that mindful participators should also hear from critics who are skeptical about the narrative of participation. When I upload my photos to Flickr.com, and tag them so my friends and I can identify them later, I'm doing myself a service: I'm exhibiting my pictures and storing them in a way that makes it easy for me to find them later. I'm also helping making Flickr communities possible. As Shirky points out, before Flickr, nobody knew that a community could form around pictures of cats in sinks (and uncounted other categories).[51]

I believe the key issue in deciding whether playbor is a fair trade of your labor for a useful service or unfair exploitation is what lawyers call "informed consent." This legal principal is why you hear those voice-over disclaimers about frightening side effects recited in television commercials for pharmaceuticals, and why your surgeon has to explain how a procedure might maim or kill you (and make you sign a paper, testifying that you've been informed and give your consent). We trust our pharmacists and surgeons with our lives, and they have a responsibility to tell us what we're getting into when they practice their services on our bodies. Informed consent online, if it exists, is buried in long privacy policies that few people actually read. And in many cases, it would seem silly for YouTube, say, to tell you all the time that you are making its owner, Google, more valuable by uploading home camera phone movies of your pets. But social media are free of visible financial charges for a reason. Knowing that reason is at the heart of informed consent online. Go ahead and improve the market value of the latest vowel-truncated social media start-up by tagging media. First, know why and how your labor is being used.

One of my education mentors, YouTube superstar and anthropology professor Michael Wesch, accurately describes the challenge to today's teachers and students (along with parents and children, and every newcomer to social media) when he observes, "We use social media in the classroom not because our students use it, but because we are afraid that social media might be *using them*—that they are using social media blindly, without recognition of the new challenges and opportunities they might create."[52] I've discussed the opportunities at some length. The challenges start with the idea that the narrative of participation I've been portraying—every smart phone as a printing press and broadcasting station, millions of cultural creators, the transformational potential of writing for a network, the gift economies of virtual communities, and fan cultures as learning networks—leaves out the ways wealthy corporations might be exploiting participatory culture. Yes, traditional monopolies on cultural creation have been contested by the proliferation of amateur videos, blog posts, and photographs. That doesn't mean that Disney, Sony, and Time Warner are powerless, or unaware of the rise of networked production.

Mirko Tobias Schäfer counters the narrative of participation in his book *Bastard Culture!* "One may ask to what extent the many user activities that were first described as a process of emancipation have been integrated into new business models and subject to corporate control. In addition, participatory culture cannot be reduced to user activity alone. . . . The specific qualities of the technology stimulate or avert certain uses and thus influence the way technologies are used and implemented by consumers in society."[53] Facebook's long history of changing the privacy settings for its users and the sensitivity of the information that search engines know about people's search histories are just two examples of specific qualities of technologies that influence the way people use Facebook or search—and effect the consequences of those activities.

Schäfer believes that the manner in which the story of participatory culture is most often told fails to include the ways commercial interests are responding to consumer-producers by evolving new means for separating people from their money and/or labor:

The computer and particularly the Internet have been represented as enabling technologies, turning consumers into users and users into producers. The unfolding online cultural production by users has been framed enthusiastically as participatory culture. But while many studies of user activities and the use of the Internet tend to romanticize emerging media practices . . . I argue that participatory culture is rather a dynamic interaction of users and companies, discourses and technologies. The availability of computers and Internet expand the traditional culture industry into the domain of users, who actively participate in cultural production, either by appropriating products from the commercial domain or by creating their own. But while user activities constitute a significant loss of control for certain sectors of traditional media industries, especially in the area of distribution, the larger culture industry benefits from user driven innovation through the appropriation of corporate design. Furthermore, the media industry undergoes a shift from creating content to providing platforms for user driven social interactions and user-generated content.[54]

Talk of empowerment, Schäfer contends, paints a utopian picture of social progress attained through technology use. He has a point: elementary crap-detection forces us to ask who might profit from such a narrative. Technology vendors and purveyors of packaged culture, perhaps? Knowing of Schäfer's critique before I started drafting this book, I was concerned enough about it to attend an academic conference in New York, the Internet as Playground and Factory, which organizer Trebor Scholz described on the conference Web site:

Only a small fraction of the more than one billion Internet users create and add videos, photos, and mini-blog posts. The rest pay attention. They leave behind innumerable traces that speak to their interests, affiliations, likes and dislikes, and desires. Large corporations then profit from this interaction by collecting and selling this data. Social participation is the oil of the digital economy. Today, communication is a mode of social production facilitated by new capitalist imperatives and it has become increasingly difficult to distinguish between play, consumption and production, life and work, labor and non-labor.[55]

A life in which these boundaries between play and work have become less distinct could be a richer world for the individual who participates in it, or it could be a subtle form of enslavement or exploitation. Consider the "ESP game," created by Louis von Ahn at Carnegie Mellon University and licensed by Google.[56] Two players, randomly matched, who can't communicate with each other, are shown the same image and given two minutes to independently submit descriptive labels. If both players use the same labels, they gain points. That's the entire game, and it was entertaining enough for players to label a hundred million images. Google licensed the game because every round improves the precision of Google's image search service. So was Google providing free entertainment, or were gamers contributing free labor to Google? Von Ahn sees a wide spectrum of possibilities for creating games that get work done as a by-product of game play. Is this a new means of exploitation or a win-win scenario? When the *Huffington Post*, a platform for unpaid bloggers, was purchased by AOL for more than three hundred million dollars, AOL began laying off professional journalists.[57] At what point does a platform for participation become a scam?

In a nontrivial sense, this book is a response to the critiques posed by Schäfer and Scholz. If these critics are correct, what is to be done? Any political or cultural countermeasures must depend on knowledge. Do people know how their labor (or play) is being used by others? That's the informed consent part. And do digital participants understand the central importance of human agency in accomplishing social progress? Technologies certainly can play a role, but the crucial thing to know is that the way people do things together via the technology is what accomplishes social progress, if any. Merely consuming the products of technology vendors is insufficient to better the human condition. Hence, my version of the narrative of participation stresses that what people know and do with their social media literacies matters. I hope that what people learn and put into action after reading this book will help them make a difference—and I want to do that by urging them to turn their crap detectors on the narrative of participatory culture as well.

Speaking of informed consent: before encouraging anybody to create digital material and attach their real name to it, I want to emphasize that our identifiable online behavior influences the way others see us. And our digital activities can influence how we think of ourselves.

Footprints and Profiles: How You Look to Others . . . and Yourself

There is some crappy stuff concerning me on the Internet—both stuff that I've produced and stuff that's been written about me by other people. It's not even a matter of blog posts or comments; I'm ashamed of some scholarly articles I've written and published! So what embarrasses me online is not just content I wrote foolishly but also content that I wrote with the intention of it being public and persistent. My way to cope with this is to constantly put up new content on the Internet that clouds out the past. To make a presence that's much more present me than past me. Can you still get to the past? Yeah, but it will take you some time. And that's part of the point—stalking me is creepy when it takes you that much effort. My way of coping with persistence is to create a living presence, frame my own story in an ongoing way, and creating a digital self that is constantly evolving not to escape but to mature.

—danah boyd, interview with the author, 2010

Life is a performance, online and face-to-face. Who doesn't behave differently in front of their parents, boss, drinking buddies, Facebook friends, or World of Warcraft guild? Decades before the Internet enabled middle-aged men to pretend to be teenage girls online, Goffman looked at the role-playing inherent in everyday interactions. Goffman's classic 1959 book *Presentation of Self in Everyday Life* shows how theatrical metaphors fit situations that everyone recognizes, such as the waiter who behaves differently "backstage" (in the kitchen) than "front stage" (with the customers). Especially useful for understanding online behavior is Goffman's assertion that people "give" information to others in order to represent who they want others to think they are—"impression management," in Goffman's terms. The way a person dresses and talks, in particular, can be carefully chosen to convey an impression intended by that person. Goffman claims that without intending to, people also "give off" information about themselves.[58] Because of the importance in human evolution of the ability to assess whether strangers are trustworthy or not, people pay attention to the information others give them and are also sensitive to the information others give off. One strong link between mindfulness and participation is the two-part question: What impression is my digital participation deliberately giving to others? And what impression is my digital participation unintentionally giving off?

One of the most powerful effects Facebook has had on a growing fraction of the human race (a half-billion users by 2011) is the way it's transforming aspects of social relationships that had heretofore been more abstract and private into a much more concrete, public form.[59] We have had friends since our species evolved, but Facebook now forces us to inscribe our friendships on our profile pages by listing our friends publicly, and publish information about our tastes in music and sexual preference, marital status, college networks, and workplaces. If we don't know how to change Facebook's privacy settings (which Facebook has notoriously changed a number of times), we disclose ourselves not only to our "friends" but also to the whole world. Not only does Facebook enable and require us to publish this information that was formerly oral and ephemeral, we cast our digital characteristics in formats that are findable through search engines, and this information is much more difficult to remove than most people are aware. As it turns out, we can learn a lot by inspecting people's Facebook profiles.

Psychologist Max Weisbuch at Tufts University recruited thirty-seven students to come into his lab and chat with another student participant.[60] One of the "students" was actually a confederate of the researchers who rated each participant's likability based on factors such as how much they smiled, their tone of voice, and their degree of self-revelation. Then ten students from another university were asked to look at the study participants' Facebook profiles and rate their likability. Those profiles that were the most highly expressive, with more pictures, videos, and wall posts, were deemed the most likable—and correlated with the participants who the research confederates had assessed as being most likable. Cornell University researchers Amy Gonzales and Jeffrey Hancock found that subjects who updated and viewed their own profiles reported greater self-esteem than those who were exposed to a mirror or an empty room—the hypothesis being that since profiles consist of people's most positive presentations of themselves, viewing the profile should increase their measurable self-esteem.[61] I cite these studies to emphasize that writing and rewriting a Facebook profile—and other forms of participation in networked publics—can influence ourselves as well as others.

Researcher boyd (no capital letters in her name), formerly a student of mine and now at Microsoft Research, notes, "Social network site profiles are where youth write themselves into being. Think of the profile as digital body. . . . It's all about self-expression for friends."[62] Although "self-expression for friends" might be more intensely significant for adolescents, it is a vital part of social media for all ages. Venture capitalist Fred Wilson, who has invested in social media enterprises such as Twitter and Tumblr, writes about the way people use Tumblr for self-expression: "A Tumblr is self expression. Jessica's looks different than Emily's, mine, and the Gotham Gal's. That's powerful. And that is what I think is driving Tumblr's popularity. Self expression matters."[63] (Emily and Gotham Gal are Wilson's daughter and wife, respectively—and Tumblr passed its one-billionth post in 2010.)[64] Tumblr differs from first-wave blogging as filtering, connecting, criticizing, or advocating by enabling people to express themselves by reblogging material they see elsewhere in a kind of collage of found social objects that reflect their vision or taste.

In addition to footprints, we create profiles—an area where participants have more potential to manage the kind of impression they give. Although understanding the potential impact of digital footprints and profiles is essential for young people who are setting out with relatively clean online slates, mindful self-presentation is important to anyone today. With well-formed digital footprints and profiles along with mindful participation in social media, you make yourself discoverable, you display your suitability to others as a collaborator, employee, or date. While you raise your positively skewed presentations of yourself, any negative, less popular representations sink lower in your search engine listings. You can't easily erase bad talk about you online; a better strategy is to dilute it with good talk. You make beneficial serendipitous encounters more likely in the cybersphere when you put thought into your footprints and profiles. Create a Google profile to take control of what the world sees when it searches for you.[65] Think about what the world sees when it searches for you; do what's in your power to do about your footprints and profiles. If you don't control your social media dossier, others will certainly use it to control you.

No discussion of digital participation can afford to ignore Twitter, the unexpected medium that ballooned to hundreds of millions of users. Twitter literacy is a perfect example of how your knowledge of how to use your attention, participate and collaborate, and use networks will determine how or whether you use the medium successfully.

Twitter Literacy

Twitter has turned out to be less an inane lifelog of what we ate for lunch and much more a streaming list of cleverly editorialized headlines with links to the main article. For many of us, Twitter is becoming the front page of our morning newspaper. Either in perception or in practice, our reporters are becoming our friends and our friends are becoming the editors of our Twitter-based newspaper.

—David Sasaki, "Our Friends Become Curators of Twitter-Based News," 2010

"The role of the internet was critical at the beginning," [Amr] Gharbeia says. "On the 25th, the movements of the protesting groups were arranged in real time through Twitter. Everyone knew where everyone else was walking and we could advise on the locations of blockades and skirmishes with police. It was real-time navigation through the city, and that's why it was shut down."

—Mike Elkin, "New Video: Cairo Geeks Survive Tahrir Square Assault," 2011

Sure, Twitter is banal and trivial, full of self-promotion and outright spam. So is the Internet. The difference between seeing Twitter as a waste of time or a powerful new community amplifier depends entirely on how you look at it—and how you grasp it. A characteristic or feature of a technology that enables a human to grasp it is known as an "affordance." A doorknob is an affordance for a door; a point-and-click command structure is an affordance for a graphical user interface. The doorknob allows a human to comprehend and manipulate the door with a hand. The point-and-click command enables a human to manipulate digital objects. Twitter has a blank box where text can be entered, a question ("what's happening?"), and a 140-character limit. Those are all affordances that both constrain and empower your use of Twitter. Twitter users can follow anyone they want (except for people who have "protected" their accounts, granting access by permission), and anyone can follow them. The following is asymmetrical. People can block others from following them. Everyone's followers are visible to others. Those are all social affordances. Taken together, those social affordances create a platform for social innovations.

A good example is the hashtag. A Twitter user, Chris Messina, thought it would be useful to create ad hoc groups of people who share an interest, a locale, or an event, whether or not they followed each other.[66] He proposed what has come to be known as the hashtag—putting the # symbol in front of a word and then using search to pull out all the tweets that use that hashtag. One of the first events to utilize the hashtag was the South by Southwest Interactive Conference, known as sxsw—so the hashtag was #sxsw. Third parties built Twitter search engines, and Twitter client companies like TweetDeck (recently purchased by Twitter), HootSuite, and Seesmic made it easy to follow all the tweets that are emitted around specific hashtags. I have TweetDeck columns displaying my Twitter lists and persistent searches for hashtags that link me to various communities.

When I started requiring digital journalism students to learn how to use Twitter, I logged onto the service and broadcast a request. "I have a classroom full of graduate students in journalism who don't know who to follow. Does anybody have a suggestion?" Within ten minutes, we had a list of journalists to follow, including one who was boarding Air Force One at that moment, joining the White House press corps accompanying the president to Africa. One of my students asked me online why I use Twitter. I replied off the top of my head. Sometimes, that's better than taking longer to compose something more elaborately thought out (which is one of the reasons I like to Twitter—it's a great way to start my word flow for the day with something short and lightweight). My other reasons are:

Openness: Anyone can join, and anyone can follow anyone else (unless they restrict access to friends who request access).

Immediacy: Twitter is a rolling present. You won't get the sense of Twitter if you just check in once a week. You need to hang out for minutes and hours, every day, to get in the groove.

Variety: The diversity on Twitter includes political or technical argument, gossip, scientific info, news flashes, poetry, social arrangements, classrooms, repartee, scholarly references, and bantering with friends. And I'm in control of deciding how much of each flavor I want in my flow. I don't have to listen to noise, but filtering it out requires attention. You are responsible for whoever else's babble you are going to direct into your awareness.

Reciprocity: People give and ask freely for information they need.

A channel to multiple publics: I'm a communicator, and have a following that I want to grow and feed. I can get the word out about a new book or video in seconds—and each of the people who follow me might also feed my memes to their own networks.

Asymmetry: The fact that nobody is required to follow those who follow them adds an interesting social twist to Twitter, because nobody sees the same sample of the Twitter population that others see. Few people follow exactly the same people who follow them. There is no social obligation to follow people simply because they follow me. I tell them that I follow people who inform or amuse me, and hope to do the same for people who follow me.

A way to meet new people: Connecting with people who share your interests has been the most powerful social driver of the Internet since day one. I follow people I don't know otherwise but who share enthusiasm for educational technology, do-it-yourself video, online activism, creativity, social media, and digital journalism—the list is as long as my list of interests. Developing the ability to know how much attention and trust to devote to someone met online is a vitally important corollary skill. PLNs are not a numbers game. They are a quality game.

A window on what is happening in multiple worlds: I am familiar with some of these worlds, and others are new to me. In chapter 5, I'll explore the power of connections to people who are unlike our most familiar communities.

Forming community: Twitter is not a community but rather an ecology in which communities can emerge. That's where the banal chitchat comes in: idle talk about news, weather, and sports is a kind of social lubrication that can enable the networks of trust as well as norms of reciprocity from which community and social capital can grow.

A platform for mass collaboration: I forgive Twestival's cute name, because this online charity event has raised over a quarter-million dollars via Twitter, funding fifty-five clean-water projects for seventeen thousand people in Ethiopia, Uganda, and India.[67] Institutions are emerging around the use of Twitter in emergency disaster relief.[68]

Searchability: The ability to follow searches for phrases like "swine flu" or "Howard Rheingold" in real time provides a kind of ambient information radar on topics that interest me. I can pull RSS feeds of those searches into my dashboard as news radars.

I still hang out on Twitter (I am found there as @hrheingold), but it's clear that many of the people I talk to about it just don't get why anyone wastes their time doing anything with the name tweeting. So I tell them that to me, the successful use of Twitter comes down to tuning and feeding. And by successful, I mean that I gain value—useful information and answers to questions along with new friends and colleagues—and that the people who follow me gain value in the form of entertainment, useful information, and some kind of ongoing weak-tie relationship with me.

To oversimplify, the successful use of Twitter depends on knowing how to tune the network of people you follow, and how to feed the network of people who follow you.

You have to tune who you follow. I mix friends who I know IRL ("in real life"), and whose whereabouts and doings interest me, people who are knowledgeable about a field that interests me, people who regularly produce URLs that prove useful, extraordinary educators, and the few who are wise or funny. I learned from master educators on Twitter that growing and tuning a PLN of authoritative sources and credible colearners is one of the success strategies in a world of digital networks. PLNs are important enough that I will detail best practices for cultivating them in chapter 5.

Feeding my network comes down to putting out the right mixture of personal tweets (while I don't really talk about what I had for lunch, the cycles of my garden, the plums falling from my tree, and my obsession with compost and shoe painting do feature in my tweet stream), informational tidbits (when I find really great URLs, that's when Twitter is truly a microblog for me to share my finds), self-promotion (when I post a new video to my YouTube channel, I share the URL—but I do not automatically post everything I blog on smartmobs.com), socializing, and answering questions. It's particularly crucial to respond to people who follow me and send @hrheingold messages to my attention. I can't always respond to every single one, but I try. I also try to be a little entertaining once in a while, when something amuses me and I think it might amuse others.

Everyone has a different mix of these elements, which is part of Twitter's charm. My followers have encouraged me to keep some personal element going, but not to overdo it. I am careful not to crank up the self-promotion too much. I don't ask questions often, yet when I do, I always get a huge payoff. I needed an authoritative guide to Spanish-language online publications about social media for a course I was designing to be taught at the (online) Open University of Catalonia. I got five. Within five minutes.

If it isn't fun, it won't be useful. If you don't put out, you don't get back. But again, you have to spend some time tuning and feeding if Twitter is going to be more than an idle amusement to you and your followers (and idle amusement is a perfectly legit use).

The use of Twitter to build PLNs, communities of practice, and tuned information radars involves more than one literacy. The business about tuning and feeding, trust and reciprocity, and social capital is a form of network literacy that I will detail in chapter 5. Knowing that Twitter is a flow, not a queue like your email in-box, to be sampled judiciously is only one part of infotention practice. My students who learn about the presentation of self and construction of identity in the psychology and sociology literature see the theories they are reading come to life on the Twitter stage every day—an essential foundation for participatory media literacy.

Attention literacy is reflective. Crap detection is analytic. Participation is deliberate. Next, I introduce the social literacies: collaboration, collective intelligence, social production, virtual communities, and other newfangled social forms that digital publics have invented—and what you need to know to be both a mindful participant and a beneficiary of cybersociality.

4 Social-Digital Know-How: The Arts and Sciences of Collective Intelligence

When we moved away from self-sufficiency and began to work together, combining our knowledge, the consequence was far-reaching: We created things we could not and do not understand, from cordless mice to urban metropolises.

Cooperation turned us into specialists: I'll do this job, you do that one. Specialization gave us incentives to innovate. Innovation led to yet more specialization and more ways of combining different specialized skills. Human intelligence became collective and cumulative to an extent that no other species can rival.

—Matt Ridley, "When Ideas Have Sex," 2011

In 1989, Tim Berners-Lee, a physicist at the European Particle Physics Laboratory, proposed "that a global hypertext space be created in which any network-accessible information could be referred to by a single 'Universal Document Identifier.'"[1] Berners-Lee, since dubbed Sir Berners-Lee, refused to patent the arrangement, which came to be known as the World Wide Web. He didn't want to own it. He wanted to use it. And he knew it would be most useful to him and other scientists if many more people used it.

By 2011, the number of indexed Web pages had grown to sixteen billion pages.[2] Berners-Lee neither had to ask for permission from any central Internet authority nor require any physical network to be rewired. The lack of a need for either permission or rewiring was possible because Berners-Lee gave away tools for building the Web on top of the existing architecture of the Internet, which had itself been designed to allow any node to propagate innovations like the World Wide Web to every other node of the network. The Web is the primary example of network-enabled collaboration on a scale that was never before possible—a phenomenon that has become known as mass collaboration. Combining the Web's built-in support for collaborative innovation (the architectures of participation mentioned by O'Reilly) with social norms of trust, sharing, and reciprocity enables people to accomplish tasks together in novel ways.

Mass collaboration has transformed not only the way people use the Internet but also how information is found (Google's PageRank), knowledge is aggregated (Wikipedia), science is conducted (citizen science), software is created (social production of the free Linux operating system and Firefox, the second most popular Web browser), computing power is harnessed for research (distributed computation), people are entertained (massive multiplayer online games), problems are solved (collective intelligence), news is gathered (citizen journalism), disaster relief is delivered (crisis mapping and emergent collective response), communities are formed (virtual communities), and commercial products are designed and tested (crowdsourcing). It isn't easy to think of a realm of human behavior that has *not* been influenced in some way by a form of mass collaboration. Although much of human culture—defined as everything we learn from and teach to one another—is transmitted through ancient institutions such as education and publishing, the techniques and social forms that made mass collaboration possible on the scale we see today have become available to billions of people in less than the span of a single human generation. The Web dates from 1989, Wikipedia dates from 2001, and Facebook dates from 2005. Although not everybody is going to crowdsource or crisis map, the knowledgeable digital citizen ought to know how virtual communities, wikis, and other varieties of mass collaboration work—and how to join in the fun.

If you tag, favorite, comment, wiki edit, curate, or blog, you are already part of the Web's collective intelligence. These online skills and the tools that enable them are part of a bigger picture: some scientists now suspect that the ability to use communication media to organize collective action might have been the force that drove primates to become humans and gave humans the leverage to create civilizations. New understandings in anthropology, biology, economics, and other disciplines have led to new theories about where humanity came from as well as new tools to address the threats humans have created to their continued existence. Mass collaboration is about doing your own business more effectively. It's also about contributing to humankind's most serious collective business of self-preservation. Faced with multiple threats from climate change, political conflicts, epidemics, and resource depletion, the most useful asset Homo sapiens has in its possession is the ability to solve problems collaboratively.

For the past ten years, in partnership with the Institute for the Future on what we called "the cooperation project," I've been helping to tell a new story about how humans get things done together. I announced the new story at the TED conference in 2005 and cotaught a seminar about it at Stanford the same year.[3, 4] It was one of the themes in my 2002 book, *Smart Mobs*. Those of us who live in urban, industrialized societies all know the old story—a narrative taught to us by our parents, schoolteachers, and (outmoded) scientific theories: biology is a war in which only the most fiercely competitive can survive; businesses and nations succeed only by dominating or destroying the competition; and politics is about your side winning at all costs. I see the outlines of a new narrative emerging from a dozen different disciplines, however. Competition is still important, but its place on our map of the universe has to shrink to make room for what we now know about cooperative arrangements and complex interdependencies in ecosystems, economies, and societies. We've compiled the results of our cooperation project work in an online knowledge base intended to catalyze an interdisciplinary understanding of cooperation.[5] Four understandings are worth keeping in mind if you want to effectively deploy the collaboration skills the Web makes possible:

• *Attention is a fundamental building block of social cooperation*, which requires paying *attention to each other*—literally the ability to look where another person is pointing. The multiple person attention dance known as learning is our species' most powerful invention.

• *Humans are supercooperators* because we've learned how to devise new tools and methods to overcome social dilemmas. These tools both augment and shape our thinking capacities by extending them beyond the limits of a single mind, and connecting them in collective intelligences and social institutions.

• *Innovative social institutions continually coevolve together with communication media*. We use the institutional frameworks we invent to modify our own behaviors in ways that build value for all. New media make possible new kinds of institutions.

• *Reciprocating cooperation, punishing noncooperators, and signaling a willingness to cooperate are useful* for individual cooperators as well as the groups they contribute to.

Humans have succeeded as a species primarily because we've learned to use our brains to cooperate in new ways, claim a growing number of evolutionary anthropologists, including Robert Boyd, Joseph Henrich, Peter Richerson, and Robin Dunbar as well as evolutionary biologists such as David Sloan Wilson. Wilson notes that simply pointing at something to attract the attention of another person is unique among primates. "Apes raised with people learn to point for things that they want but never point to call the attention of their human caretakers to objects of mutual interest, something that human infants start doing around their first birthday," writes Wilson.[6] Infotention. "Watch what I'm paying attention to" is the elementary particle of cooperation.

Dunbar also studies communication among primates, particularly the behavior known as social grooming. Empirical evidence indicates that picking lice out of each other's hides, a primitive but direct form of social reciprocation, enables bands of primates to develop the bonds of trust necessary to compete successfully with other bands. Dunbar compared measurements of the size of social groups among thirty-six different primates, and found a direct correlation between the size of the group and the brain size of group members. The title of Dunbar's seminal 1993 paper summarizes his hypothesis: "Coevolution of Neocortex Size, Group Size, and Language in Humans." Human higher brain functions, Dunbar asserts, evolved in order to process social information such as remembering faces, keeping track of histories of interactions, recognizing who to reward and who to punish, and especially knowing who to trust and not to trust. According to Dunbar, gossip was the first use of language—and one that has not gone away: "Analysis of a sample of human conversations shows that about 60% of time is spent gossiping about relationships and personal experiences. It is suggested that language evolved to allow individuals to learn about the behavioural characteristics of other group members more rapidly than is possible by direct observation alone."[7]

Dunbar also claims that human neural capabilities both enlarge and limit the size of our social groups. He maintains that "there is a cognitive limit to the number of individuals with whom any one person can maintain stable relationships, that this limit is a direct function of relative neocortex size, and that this in turn limits group size. . . . [T]he limit imposed by neocortical processing capacity is simply on the number of individuals with whom a stable inter-personal relationship can be maintained." Extrapolating from the thirty-six other primates, Dunbar predicted that the mean group size of a socially stable group of humans is 147.8—a figure that is borne out by census data of village and tribal societies, and also happens to be approximately the size of basic military units.[8] (I'll revisit "Dunbar's number" in the next chapter, when I take up social networks.)

Where Sloan Wilson looked at the biologically evolved capacities of humans and Dunbar studied the origins of language, Boyd, Henrich, and Richerson examined the evolution of sociality as a cultural phenomenon. A hundred thousand years ago, our ancestors were small, slow, and lacking in claws, fangs, or wings in a hostile environment, but they used the social and information-processing capacities of their brains to organize collective defense and food gathering. Innate human propensities for cooperation with strangers, shaped in response to rapidly changing environments in the Pleistocene, could have provided highly adaptive "social instincts."

Although the biological capacity for primate sociality evolved genetically over millions of years, Boyd, Henrich, and Richerson as well as others propose that the channeling of behavior via symbol systems has catapulted the evolution of cooperative human capacities from the slow biological level to the faster-paced one of cultural change.[9]

The social instincts were biological, but they made it possible to influence behavior through human-invented institutions. The invention of culture—symbolic communication, learning by imitation (which requires paying close attention to others), and cooperative methodologies such as dance and ritual—leveraged social instincts to deliberately promote human cooperative advantages. Institutions—social arrangements that persist over time and channel the group behavior—matter. A swarm of human social and cultural inventions self-orchestrate to create this communication-cooperation-innovation system that has propelled us from savanna to high-rise, including "punishment, language, technology, individual intelligence and inventiveness, ready establishment of reciprocal arrangements, prestige systems, and solutions to games of coordination."[10] Punishing those who break the institution's rules is apparently essential to cultivating cooperation; "altruistic punishment" may be the glue that holds societies together.[11] Seeing humans as supercooperators who constantly seek to invent ways to increase their abilities to cooperate makes visible an entirely different landscape of possibility than can be revealed through the story that focuses solely on humans as ruthlessly successful competitors. Social media are part of a lineage that connects cave paintings to alphabets to hyperlinks. Virtual communities, like contracts and constitutions, are technologies of cooperation.

The institutions that have been pivotal in the acceleration of cooperation from the biological to the cultural level are solutions to a set of problems known as "social dilemmas." Social dilemmas are the conflicts between self-interest and collective action that all creatures face in daily life—situations in which a lack of trust in the potential cooperation of others prevents individuals from acting together in ways that would benefit everybody. This tension exists at every level from the cellular to the social, but takes on particular significance with human transactions, which can include such symbolic currencies as money, reputation, social capital, and a sense of identity. Social dilemmas arise over the consumption and provision of resources. Fishing licenses are an example of an institution that attempts to solve a common resource dilemma: if everybody gets to fish as much as each desires, then the stock will be overfished and disappear, as happened with Atlantic cod.

The social dilemma known as the "tragedy of the commons" became a kind of mythical narrative when biologist Garrett Hardin wrote a paper with that title in 1968, warning of the dangers of human population growth. He referred to the fate of the common fields that became overgrazed when individual farmers, unrestrained by state regulation or private property rights, each added to their flocks until the meadows they shared became unusable. The tragedy that Hardin feared was the inevitability of human self-interest causing disastrous problems for collective interests—unless commons are either privatized or strictly regulated by the state.[12]

Years after this controversial essay, political scientist Elinor Ostrom asked whether empirical data support Hardin's conclusion that humans will invariably despoil any common resource. Looking at thousands of records, ancient and modern, of the human use of shared watersheds, fishing and hunting grounds, and forests and grazing lands, Ostrom discovered that a significant portion of communities found ways to override basic social dilemmas by constructing systems of norms and self-policing social contracts. The tragedy of the commons is common though not inevitable—but only if people deliberately invent a social work-around. Ostrom (who won the 2010 Nobel Prize in economics for this work) found that these work-arounds, which she called "institutions of collective action," were more likely to succeed when a small number of design principles were observed, and more likely to fail in the absence of these measures. She also observed that groups that are able to organize and govern their behavior successfully are marked by the following design principles:

1. Group boundaries are clearly defined.
2. Rules governing the use of collective goods are well matched to local needs and conditions.
3. Most individuals affected by these rules can participate in modifying the rules.
4. The right of community members to devise their own rules is respected by external authorities.
5. A system for monitoring member's behavior exists; the community members themselves undertake this monitoring.
6. A graduated system of sanctions is used.
7. Community members have access to low-cost conflict resolution mechanisms.
8. For common pool resources that are parts of larger systems: appropriation, provision, monitoring, enforcement, conflict resolution, and governance activities are organized in multiple layers of nested enterprises.[13]

Other kinds of social dilemmas offer obstacles to commerce, and thus provide a motivation to solve them and make a profit. An unsecured transaction, for example, is a dilemma for both the potential buyer and potential seller: someone you don't know in another city will send you something you want at a good price; you don't want to send the money until you get the goods, and the seller doesn't want to send the goods until you deliver the money. Trust lubricates markets. A lack of trust prevents them. The online auction site eBay solved this social dilemma by enabling buyers to give public ratings to sellers after transactions have been completed; information available to potential future buyers about a seller's past history and reputation furnishes the amount of trust necessary for a multibillion dollar market to exist where none had been possible before.

From cave paintings to Wikipedia, the power of networked media often stimulates new ways to share and act together. When people choose to invest in a market, raise an army, start a religion, or support a political party, sociologists label this behavior "collective action." The more commonly used words coordination, cooperation, and collaboration differentiate important aspects of collective action. Knowing the difference between these terms is the place to sharpen your understanding of both cooperation and collaboration. When I broadcast the question on Twitter to my network of followers, What's the difference between these three related but not identical forms of collective action? a professor in Canada, Wayne Macphail, put it this way, two minutes after I tweeted my query: "You need coordination to dance, cooperation to dance with a partner, and collaboration to dance with a flash mob." When you send a text message to your friends to meet you at a restaurant, you are coordinating. When you contribute your computer's idle computing power to medical research, you are cooperating. And when you and three friends use a wiki to plan a road trip or revolution, you are collaborating.

By discussing my interest in defining collaboration in public via Twitter, I attracted the counsel of someone I've learned to rely on as a trusted source on matters of collaboration, Eugene E. Kim. I've found that the right kind of "thinking aloud" in places that my networks gather frequently saves me hours or days of work with a few simple pointers (pointing is important) from my networks. Kim responded to my public questions by saying that his own work with wiki collaboration was based on the work of Arthur Himmelman, who detailed the differences between networking, coordination, cooperation, and collaboration. Himmelman's taxonomy makes a great deal of sense to me. Here, paraphrased in my own words, is how Himmelman differentiates these four related components:

1. *Networking* is the simplest, with the least risk and commitment—such as handing out business cards, attending conferences, hanging out in a chat room, or commenting on a blog.

2. *Coordination* means that all involved parties share information and agree to modify their activities for mutual benefit. Neighbors need to coordinate the time they flood rice paddies in order to control the population of vermin who eat the crop. If all the neighbors don't flood at the same time, then the vermin will simply move to the nearest dry ground and threaten the whole neighborhood come harvesttime. Coordination requires more commitment than networking, but not as much as the next step: cooperation.

3. *Cooperation,* as Himmelman defines it, is "exchanging information, modifying activities, and sharing resources for mutual benefit and to achieve a common purpose."[14] The amount of commitment and risk is higher than coordination (although cooperation often involves coordination), and the move from "mutual benefit" to "common purpose" requires moving from self-interest (each involved party can judge its own benefit) to agreement on what, exactly, all parties agree to hold as common goals. Sharing resources is a big step. Modifying activities may involve more systemic change by each party than simple coordination requires. Himmelman emphasizes that groups can move from networking to coordination to cooperation by building trust, communicating, and explicitly seeking a common purpose. Hunting small fish and game with relatives didn't bring in much protein for humans, but when families began to band together in larger groups of nonrelated people to hunt big game and trap fish collectively, they harvested protein bonanzas. Ostrom's institutions for collective action are social instruments for managing cooperation. The various kinds of online-mediated collective action I group together as forms of mass collaboration involve some kind of cooperation. In some but not all circumstances, these practices involve collaboration, as Himmelman defined it. Virtual communities, for example, are cooperative, but only become collaborative when they focus on a shared goal.

4. *Collaboration* is the most purposeful means of collective action. It uses networking, coordination, and cooperation as building blocks, adding to "exchanging information, modifying activities, etc.," the requirements of "enhancing the capacity of another for mutual benefit and to achieve a common purpose by sharing risks, resources, responsibilities, and rewards."[15] People collaborate because their coordination, sharing, and attention to common goals creates something that none of the collaborating parties could have benefited from without collaboration. Collaborators develop and agree on common goals, share responsibility and work together to achieve those goals, and contribute resources to the effort.

Keep a few ideas gleaned from cooperation theory in mind when dealing with the practical circumstances of life online in a digitally networked world:

What Cooperation Theory Teaches Us about Life Online Today

1. *Balance retribution and forgiveness.* As your first visible move in a new interaction with another person or group, always present yourself as ready to cooperate. If the other party proves trustworthy by mutually cooperating in response to your move, you've both started out ahead of the game for any future interactions. If the other party proves untrustworthy at the beginning, withdraw your trust and fail to cooperate on the next move (or look for new partners). If the other party changes strategy in response to your action and starts to cooperate, then switch back to cooperation. This strategy, known as "tit for tat," has proven to dominate other strategies in social dilemma experiments. It doesn't work in all circumstances, but it can be handy online.

2. *Contribute publicly* without requiring or expecting any direct reward. Signaling your willingness to cooperate can draw potential partners to you. And you'll learn by doing.

3. *Reciprocate* when someone or some group does you a favor.

4. *Look for ways to seek a sense of shared group identity* when encountering strangers. Look for commonalities, and then emphasize them.

5. *Introduce people and networks to each other* in mutually beneficial ways. Encourage interdependence.

6. *When progress is blocked by social dilemmas, create institutions* for collective action. Pay attention to Ostrom's design principles.

7. *Punish cheating, but not too drastically* (Ernst Fehr and Simon Gächter's altruistic punishment meets Ostrom's graduated sanctions), and enforce punishment through community-wide social disapproval, rather than officially policed laws, whenever possible.

What We Can Learn from Collaboration Theory about Life Online

1. *Small talk and idle chatter build trust* and lubricate collaboration. If you want your engineers to collaborate, don't shut down their beer-brewing email list. People talk about sports in order to start a conversation that looks for commonalities and probes for trustworthiness. The casual sharing of experiences on Facebook and Twitter that some dismiss as trivia can serve a purpose if it enables trust building.

2. *Move from mutual benefit to common interests* by building trust and negotiating goals.

3. *Take risks* to demonstrate that you are willing to modify your own activity in pursuit of common goals.

4. *Be generous* by sometimes giving away what you might have wanted to sell or sequester. Always feed networks without seeking direct reciprocation, especially if the cost to you is low.

5. *Seek to learn from and teach your collaborators.* Be willing to change your behavior in light of learning, and be willing to help your partners enhance their own positions.

Some of the most exciting new literacies are emerging from the intensive collaboration laboratories known as massively multiplayer games, where the ancient arts of collaboration meet modern technologies of cooperation. Jane McGonigal, director of game research and development at the Institute for the Future, estimates that about three billion hours a week are spent playing computer and video games. She also believes that some gamers are learning to be supercollaborators. I take her seriously.

McGonigal created a collaborative game for the International Olympic Committee that attracted a quarter-million gamers from more than one hundred countries. When I interviewed McGonigal at the Institute's headquarters in Palo Alto, she told me about the Lost Ring, the game sponsored by McDonald's for the International Olympic Committee. One of the first things she said was that evoking a sense of "awe and wonder," of participating in an overarching ("epic," in gamer speak) cause, is essential to conjuring mass collaboration from strangers online.[16] So the Lost Ring started with a series of podcasts by a fictional character who talked about hidden artifacts and secret codes in six languages, with physical clues scattered around the world, all regarding a legendary "lost Olympic game." Drawn by links on the official Olympics Web site, gamers created a 943-page multimedia wiki in six languages to collaborate on finding the physical clues, breaking the codes, and communicating across language as well as geography. Not only did volunteer teams around the world decrypt the rules of the lost game, they formed teams and practiced it—a runner tries to traverse a physical labyrinth while blindfolded, guided only by the humming sounds made by teammates—and uploaded hundreds of videos of their performances. Just search YouTube for the Lost Sport.

In her book *Reality Is Broken: Why Games Make Us Better and How They Can Change the World*, McGonigal writes: "We are all born with the potential to develop collaboration superpowers. Scientific research shows that we have both the ability and the desire from early childhood to cooperate, to coordinate activity, and to strengthen group bonds—in other words, to make a good game together. But this potential can be lost if we don't expend enough effort practicing collaboration."[17] In her talk at the influential TED conference, McGonigal noted that after Wikipedia, the "second biggest wiki in the world, with nearly 80,000 articles is the WoW [World of Warcraft] wiki. [Five] million people use it every month. They have compiled more information about WoW on the internet than any other topic on any other wiki in the world."[18]

McGonigal has developed a model of how collaboration superpowers work:

Extraordinary collaborators have no qualms about **pinging**—or reaching out via electronic means—to others to ask for their participation. . . . Of course, it helps to have a good sensibility about who to ping when. . . . That's why extraordinary collaborators develop a kind of internal **collaboration radar**, or sixth sense, about who would make the best collaborators on a particular task or mission. This sixth sense comes from building up a very strong social network and maintaining a kind of peripheral awareness of what other people are doing, where they are, and what they're getting good at. . . . Finally, the most extraordinary collaborators in the world exercise a superpower I call **emergensight**. It's the ability to thrive in a chaotic collaborative environment.[19]

McGonigal's collaboration superpowers would work well within Ostrom's design framework for collective action institutions. What Ito calls hanging out through Twitter, instant messaging, and chat, as mentioned earlier, helps like-minded strangers create the ambient sociality from which more robust forms of collaboration emerge.

McGonigal cited several games designed by others that are already harnessing gamer collaborative superpowers to do good in the world. Foldit helps biomedical researchers understand how protein molecules fold by presenting unfolded proteins to game players as puzzles—attracting fifty thousand participants.[20] When British politicians were revealed to have cashed in millions of pounds of bogus expense receipts, Parliament made its members' receipts public, but in an unmanageably large collection of unordered documents. The UK *Guardian* newspaper created an online game that presented players with randomly chosen receipts. When players found receipts they considered questionable, they could click an "Investigate This" link. Those players with the most hits were listed on a leader board. In less than four days, twenty thousand players investigated 170,000 receipts—a brilliant and unexpected combination of game mechanics, mass collaboration, and citizen journalism.[21]

If you've ever witnessed or participated in the passion of hard-core collaborative games, you understand why teachers would love to see their students apply themselves with equivalent zeal to their curriculum, and why managers would love to see business tasks attacked with the collective vigor that teams exhibit on fantasy quests in virtual worlds. One of the most astonishing statistics McGonigal cites in her book is the estimate that gamers have spent 5.93 million years playing World of Warcraft.[22]

Individual players choose roles that have certain capabilities and powers, then seek to improve their in-game identity—their avatar—by working out puzzles, quests, and battles that bring strength, magic, wisdom, and other powers to their warrior, wizard, or gnome. But players can't get far or have much fun individually. Eventually, each World of Warcraft player joins a guild that self-organizes dozens of players to advance in the game by collaboratively succeeding in various raids. Each guild must coordinate the right mix of skills, talents, and experiences as well as execute a coordinated plan in order to succeed. "This process brings about a profound shift in how they perceive and react to the world around them. They become more flexible in their thinking and more sensitive to social cues," claim Douglas Thomas and John Seely Brown in "You Play World of Warcraft? You're Hired!"[23] Indeed, people have been meeting physical world collaborators, hiring employees, and even establishing relationships with investors in World of Warcraft often enough for the word to get around in Silicon Valley that "World of Warcraft is the new golf." It all has to do with the way raiding in a World of Warcraft guild requires what those who talk about gaming and learning frequently call "the collaborative mind-set."

I asked my friend Joi Ito about how a gamer's attitude might possibly become useful in the real world. I first met Ito through a tiny convocation of online communication enthusiasts in 1986 (and through Joi, I met his sister, Mizuko, who became my guide to the world of hanging out, messing around, and geeking out). In 1993, Joi was the first person to show me his personal home page on the World Wide Web. He's been a venture capitalist for the last decade—and known among the digerati as a hard-core World of Warcraft player and guild leader. In April, 2011, Ito was selected to be the director of MIT Media Lab, despite his lack of any college degree. "If you've never read any business case histories, but if you've run a guild, or organized a raid, or spent time resolving drama and disputes in World of Warcraft, your mind-set is well prepared for the real world in a very different way than a college MBA would be prepared to run a company," Ito told me when I interviewed him one summer afternoon in 2010, under my plum tree. He had his laptop open to his World of Warcraft dashboard while we were talking.[24]

Knowing a bit about evolutionary psychology, the sociology of collective action, and theories of game mechanics can prepare your mind-set, but only participation and direct experience can truly hone online collaboration literacies. When you move from knowledge about cooperation and collaboration to the actual practice of online collective action, I emphasize the power and skills necessary to participate in five specific forms of mass collaboration. Without a doubt, new technologies, new media, and new social practices will spawn new forms. Today, the most important include collective intelligence, virtual communities, social production, crowdsourcing, and wiki collaboration.

Collective Intelligence: When All of Us Can Be Smarter Than Any of Us

Wikipedia, as I discuss in *Convergence Culture*, depends on what Pierre Lévy calls "collective intelligence." In the classic formulation, collective intelligence refers to a situation where nobody knows everything, everyone knows something, and what any given member knows is accessible to any other member upon request on an ad hoc basis. Lévy is arguing that a networked culture gives rise to new structures of power which stem from the ability of diverse groups of people to pool knowledge, collaborate through research, debate interpretations, and through such a collaborative process, refine their understanding of the world.
—Henry Jenkins, "Collective Intelligence vs. the Wisdom of Crowds," 2006

Collective intelligence comes in many flavors. A tribe of television fans known as "spoilers" mobilize a kind of collective intelligence on a global scale in order to outwit the secretive producers of programs such as *Survivor* by figuring out who is going to win before the official announcement. Because the *Survivor* series is recorded months before it is broadcast, with the cast and crew sworn to secrecy, only a few people know who the final surviving winner is before the announcement. By talking to bartenders in hotels where the cast might have stayed, piecing together small pieces of information contributed by members of the online spoilers community, and arguing vociferously online about what the evidence means, spoiler communities have rivaled intelligence services in their ability to collectively sleuth out the answers.

Jenkins, McGonigal, Brown, and the Ito siblings believe that what might appear to be esoteric practices of digital subcultures might also be harbingers of the kind of networked problem solving that people will be doing more often in the face of more serious objectives. Wikipedia, which I'll look at shortly, is certainly the product of a culturally evolved collective intelligence that integrates the large and small efforts of many, resolves editing disputes, and maintains self-healing procedures to protect against vandalism. With the aid of crowdsourced citizen data gatherers, some scientists think that science—one of the most potent forms of collective intelligence—could be accomplished through mass collaboration. The groups of bloggers who investigated and eventually caused the resignations of Lott and Rather, the bloggers who organized the Sinclair Broadcast Group boycott, and the reconstructors of the Lost Ring fictional Olympic game all exhibited characteristics of collective intelligence—a networked rather than hierarchical command-and-control structure; a means of fitting together small contributions of many participants utilizing intracommunity forums for discussing, sharing, and arguing about their collaboration along with what it means.

In the late 1990s, University of Ottawa professor Pierre Lévy built on existing ideas of "global brains" and "mass minds" by H. G. Wells, Howard Bloom, and others in his book *Collective Intelligence: Mankind's Emerging World in Cyberspace*.[25, 26, 27] Lévy sees collective intelligence as an aggregation of skills, understanding, and knowledge: skills, like literacies, unlock individual access, which must be actively used to acquire individual knowledge from the great collective pool amassed through literate culture; and understanding requires communication, discussion, debate, and direct experience (which Socrates claimed couldn't be conveyed through unliving texts). Lévy shares my view that the Web's advent accelerates and broadens the scope of a process of bootstrapping intellectual tool sets that has been ramping up since the Sumerians invented writing. Now that we have gained access to digital tools that enable us to share what we know and aggregate small contributions into large knowledge repositories, a new level of collective intelligence is possible, Lévy asserts.

I interviewed Lévy about the skills needed to participate in and instigate collective intelligence activity today. "The essence of this new skill," Lévy told me,

is to create a synergy between personal knowledge management and collective knowledge management. You have to connect to people and find information sources, then filter, select, and categorize information for your own purposes. You have to decide which information to accumulate personally, to store or memorize. When you do this, you can share your personal knowledge with knowledge communities through social bookmarking or blogging or Twitter. When you tweet a URL, you usually include a brief commentary. The comments you share should help people categorize the knowledge you are signaling.[28]

Add value to information you find—help others transform it into knowledge by adding context—and share not only what you find but also what you think of it. The participatory skill of curation and some of the collaborative skills of virtual community are fundamental to collective intelligence.

Thomas Malone, professor of management at the MIT Sloan School and founding director of the MIT Center for Collective Intelligence, not only strives to discover and apply knowledge about augmented collective intelligence to the challenges of business management but also to some of those "complex, urgent" problems that Engelbart and others foresaw—such as climate change. One of the center's projects is Climate CoLab, an online site that "seeks to harness the collective intelligence of thousands of people around the world to address global climate change. Inspired by systems like Wikipedia and Linux, the Climate CoLab is a global, on-line forum in which people can create, analyze, and select detailed proposals for what to do about climate change."[29] The CoLab combines three tools: *open modeling* involves creating simulations of the effects of proposed solutions; *large-scale argumentation* structures discussion around issues, positions, and arguments; *group decision making* uses voting and rating systems to select the most convincing proposals. The center focuses on the practical hows and whys of collaboration in contemporary institutions, including but not limited to business management; it is also scientifically focused on measurable aspects of collective intelligence. Malone was one of the authors of a multiauthor study that reported to the prestigious journal *Science* in 2010 on "evidence for a collective intelligence factor in the performance of human groups."[30]

Psychologists have confirmed the value of a single statistical measure called "general intelligence," which can be derived from individual performance on a range of cognitive tasks (IQ) and can be used to predict performance on other cognitive tasks. The Center for Collective Intelligence research group studied 699 individuals in groups of two to five, measuring collective performance on puzzles, negotiations, judgments, and brainstorming exercises, finding that a measure of a group's collective equivalent to general intelligence can predict that group's performance on more general tasks such as playing checkers against a computer. The research team discovered that the group's collective IQs were *not* correlated with the average IQ of the individual group members, nor with the score of the "smartest" person on the team. The factors found to contribute to a group's collective IQ are the group's facility at taking turns in conversations, members' sensitivity to social cues, and the number of women on each team. Certainly the finding of a gendered element to collaboration is likely to stimulate further research. The science of collective intelligence is young, but a few cues can be gleaned from what is currently known:

Collective Intelligence Tips

1. *Encourage casual conversation* about off-topic matters. It helps build networks of trust.

2. *Diversify your group.* Recent research has demonstrated that the diversity of a social network is the most important factor influencing whether or not a person can use social capital developed within that network. (More on that in the next chapter.) Don't invite only experts, and above all, don't forget to include women.

3. *Practice collaborating* among yourselves. Learning how to distribute conversational turn taking can pay off.

4. *Make it easy to contribute* to a shared knowledge repository.

Not all collective intelligences are communities, nor does collective IQ play a major role in most communities, but the arts and sciences of virtual community represent the collaboration literacy that most digital citizens are going to encounter in their lives online.

Virtual Communities: Networking Hearts as well as Minds

A virtual community is a group of people who may or may not meet one another face to face, and who exchange words and ideas through the mediation of computer bulletin boards and networks. Like any other community, it is also a collection of people who adhere to a certain (loose) social contract, and who share certain (eclectic) interests.

—Howard Rheingold, "Virtual Communities," 1987

On more than one occasion, I've sat in vigil and companionship by the bedside of a dying member of our virtual community. I found myself on the receiving end of moral and physical support from people around the world, many of whom I had never met face-to-face, when I experienced my own health crisis. An online group I started in 1988, and which still continues, has distributed more than twenty thousand dollars over the years to members who needed to pay their rent, feed their children, and finance their surgery. I've danced at weddings of people who met in these online gatherings, babysat for their children, and attended their graduations and funerals. I've found paying work and also hired people through these groups. I've shared meals or drinks in London, Tokyo, Istanbul, Amsterdam, Memphis, and two dozen other cities with people I initially met and came to know only through computer-mediated communications. Don't tell me that my life online isn't real. If you agree with sociologist Barry Wellman's definition that "communities are networks of interpersonal ties that provide sociability, support, information, a sense of belonging and social identity," I still stand by the term virtual community as it (sometimes) applies to computer-mediated relationships.[31]

I must add that networks are not the same as communities, although any person can belong to both networks and communities simultaneously: people can communicate online about shared interests without establishing personal relationships. Knowing the difference between a community and a network is as critical socially as crap detection is essential informationally. And knowing how to be a good community or network member can be more than a key to sociability, support, information, and communications; it can save your life.

To me, the difference between an online social network and a community has to do with the quality, continuity, and degree of commitment in the relationships between members. This comes down to whether participants care about each other and are willing to act on their feelings. If I didn't show up online for a while, would anyone knock on my physical door to see if I'm OK? Community is not about always getting along well. Indeed, established norms for working through interpersonal conflicts are one of the hallmarks of communities. And although much of the time spent experiencing virtual community comes under the general heading of hanging out—talking casually about things both deep and shallow—virtual communities can organize collective action in the physical world—for example, the Harry Potter Alliance, an online fan community, financed, procured, and transported more than eighty thousand pounds of critical supplies to Haiti after the 2010 earthquake, and participants in the EVE Online fantasy and role-playing community organized more than sixty thousand U.S. dollars in relief funds in the days after the Japanese tsunami of 2011.[32, 33]

If there's one literacy I'm qualified to talk about from hard-earned personal experience, it's what the old-timers would have called netiquette—the norms of behavior that were once propagated among the Internet pioneers in order to make life less stressful and more productive for everybody. The first norm was: *Pay attention before you join in.* Sample the stream of comments daily for a few days to get a sense of a blog community, chat room, and discussion board. When I raised the subject of virtual community literacy with Jenkins, he responded,

I think the first action to take is to try to understand what kind of community you are participating in and what its norms are. The white paper we wrote for MacArthur described the process of negotiation where members of the community understand the expectations of that community and behave appropriately in that space. A construct of fictional identity is totally acceptable in World of Warcraft, but may be less desirable on something like Facebook. The disclosure of certain kinds of information may be obligatory in some communities, and objectionable or shocking or out of place in others. So the first step in being a digital citizen is figuring out what kind of world you are engaging. There are many Web communities, each with different norms of communication and participation, different levels of trust, different levels of intimacy, different expectations about what it is to be a member, and this can be very perplexing.[34]

Observe the way people interact in the community that interests you. If you have a question, ask someone quietly (that is, avoid broadcasting your question to the entire community if possible). Any community that rebuffs the informed question of a newcomer who has looked around and understands the local mores is not a community worth investing yourself in.

Without tone of voice, facial expression, and body language, text-only online discussions strip a surprising amount of emotional context from the cues we use to surmise what other people really intend. It's easy to mistake mild sarcasm online as a personal attack, leading to what old-timers called flame wars, and that abound to this day in social media forums of all kinds. When I started my third or fourth virtual community, one of the first things I began asking all members to agree to is to *assume goodwill.* If it seems to you that someone else is directing a negative communication your way, I advised newcomers, assume that the lack of social cues is causing you to misinterpret: ask friendly questions to find out what the other party intended.

A third useful tool for aspiring virtual communitarians is to *jump in where you can add value.* I adopted this attitude from the Wikipedians, some of whom write scholarly articles, and some of whom clean up misplaced commas. When introducing my college students to wiki collaboration, I tell them:

One of the coolest aspects of collaborative work like collective wiki editing is also hard to recognize at first. You don't have to do something perfectly, you don't have to make the kind of effort required for putting together an entire document on your own; rather, you look for something, anything, you can do that can add a little bit of value. A student in this course put it well in a blog post when she wrote: "I simply need to make a habit of being active in this learning community, because instead of just jumping in and participating where I can add value, I've been psyching myself out." Exactly. Just do something. If every one of us does one small thing, the aggregate of all the small things we do adds up to something significant. After you do one small thing, I bet you'll feel like doing something slightly larger. There's a certain pleasure in creating something together. Why else do people contribute millions of entries and billions of edits to Wikipedia?

Finally, *reciprocate* when someone does you a favor or shows a courtesy. If you are unknown to the community, reciprocate in advance. Pay it forward by answering questions or responding in other ways to the needs of community members. Show your willingness to help others. As we'll see when I examine social capital, there is hard evidence that the strongest predictor of whether someone will assist you online is whether you've aided others.

Do those four things—know the territory, assume goodwill, jump in wherever you can add value, and reciprocate—and you'll succeed as a virtual community member.

Now that it's easy to start a mail group, publish a blog, and create a chat room or wiki with a few keystrokes, knowing something about the art of hosting successful online communities is a skill that millions of people find useful. In 1998, I published online (we'd now say that I blogged) a now much-linked essay titled "The Art of Hosting Good Conversations Online."[35] To summarize its argument, an online host wants to achieve a feeling of ownership by the group, whereby participants become evangelists; facilitate a spirit of group creativity, experimentation, exploration, and goodwill; and encourage a shared commitment to work together toward better communication and conversations. Good online discussions help people make contact with other people, entertain themselves rather than being only passive consumers of entertainment created by others, create gift economies for knowledge sharing, and establish conditions for ongoing collaboration that reward individual effort with a whole that is greater than the sum of its parts; they also make newcomers feel welcomed and contributors feel valued, while ignoring those who hassle others as their recreation.

A virtual community organizer is like a party host. You don't automatically throw a great party by hiring a room and buying some beer. Someone needs to invite an interesting mix of people, greet them at the door, make introductions, start conversations, avert fisticuffs, and encourage people to let their hair down and entertain each other. Good hosts model the behavior they want others to emulate: read carefully and post entertainingly, informatively, and economically, acknowledge other people by name, assume benevolence, assert trust until convinced otherwise, add knowledge, offer help, be slow to anger, apologize when wrong, politely ask for clarification, and exercise patience when tempers flare. They nurture the community memory, pointing newcomers to archives and explaining inside references, providing links to related conversations past and present, and hunting down resources to add to the collective pool of knowledge— and teaching others to do the same.

Communities don't just happen automatically when you supply communication tools; online communities grow under the right conditions. They are gardened. Online social systems will not cohere without careful intervention. Build one, and nobody will come, and if they come, they won't necessarily start a self-sustaining conversation. But the intervention has to be ground up, not top down. Positive effort is required to create the circumstances and tend the growth of a self-sustaining group.

A small number of simple, clear rules, sparsely enforced, with an explicit expectation that the community's own norms will emerge later, are important at first. Establish the rules at the outset and then move on. Those who don't like it will leave. The rest will make up their own minds after they get to know each other and the system. Making rules after launching or changing them from the top down is a recipe for a community-killing rules thrash. Provide a decision-making protocol at the beginning; don't negotiate it online or you will spiral into endless, toxic metacommentary about how to decide to decide.

Eventually, natural hosts emerge in each community, and existing hosts should scout and mentor them. In regard to host behavior, patience is rule numbers one through three. Deliberately add a time delay on your emotional responses before you make any public posting or send a private email. In most cases, *not* saying anything is the best decision. The first art of the host is the knowledge of how and when not to act. Bend over backward to be fair and civil when challenged. You are performing the public drama of the community's foundation myth. Have fun! Signal that it's OK to experiment as well as not take yourself and the whole enterprise too seriously. One ounce of elegance and grace is worth ten pounds of argument. You can charm or seduce discussions back on topic, and steer conflicts away from the brink of brawl, but you can't force them. Force backfires on authority online. You have to persuade and pull because pushing is an automatic loss for authority. People's first reactions are most important. Praise them by name. Be interested. Read their profiles and point them to information that you think will be personally relevant to them. Communicate privately via email with both promising newcomers and troublemakers. Never disclose private email in a public forum. Point out the pitfalls of the medium that cause people to misunderstand each other.

I've found the skills of participating in virtual communities along with creating my own virtual communities to prove valuable in finding friends, having fun, learning what I need to know, seeking support during a medical crisis, organizing business and political activities in the physical world, finding and organizing collaborators for ventures lasting a week or decade, writing books, and growing learning communities and PLNs. These skills aren't going away, even if we now call it social media instead of virtual community.

Crowdsourcing: Supersizing Cooperation

However, crowdsourcing is not directed towards other organizations but directed towards the crowd in the form of an open call. This presents a crucial characteristic of crowdsourcing which differentiates it from outsourcing and subsequently emphasizes the need of participation of the crowd in the crowdsourcing initiative for its success. In crowdsourcing, the participation is voluntary and the contribution of a wide network of people is required for the initiative to reach a substantial scale. Therefore, sufficient crowd participation is imperative for the success of a crowdsourcing initiative. In order to reach the critical mass in terms of crowd participation the incentive ought to be tailored to attract the most effective collaborators and the motive of the crowd needs to be aligned with the long term objective of the crowdsourcing initiative. This ensures that the crowd is willing to participate in the initiative which involves either completing micro tasks or contributing information for future use. It also brings to light the importance of acceptance of the concept of crowdsourcing by the crowd (sometimes referred to as users in the paper) for the initiative to become successful.
—Ankit Sharma, "Crowdsourcing Critical Success Factor Model," 2010

On January 28, 2007, Jim Gray, a sixty-three-year-old computer scientist and experienced sailor took his forty-foot cruiser out of San Francisco Bay on a mission to scatter his mother's ashes at sea. He never returned. No radio calls or emergency beacons signaled his disappearance. Gray had made significant discoveries in the field of distributed computation and had many friends in the computer science world—some of who mobilized extraordinary resources to try to find their friend. Google and NASA provided recent satellite photographs of the forty thousand square miles where Gray had gone missing. Amazon engineers divided the photographs into a half-million fragments and used Amazon's Mechanical Turk engine to make these images available to twelve thousand online volunteers. This unprecedented mobilization of resources failed to find Gray, but illustrates just one aspect of the multifaceted power of crowdsourcing: tackling massive problems by dividing them among online volunteers and aggregating their efforts.

Jeff Howe, writing in *Wired* magazine, gave the name crowdsourcing to the phenomenon of breaking problems or tasks into small pieces, and then making an open call for voluntary participation.[36] One of Howe's examples was Innocentive, the "solution market" that the pharmaceutical company Eli Lilly created in 2001 in order to take advantage of brainpower outside the company—a daring move to crowdsource ideas in a notoriously secretive industry.[37] Scientists with problems to solve ("seekers") can pose them in the marketplace. Others ("solvers"), not necessarily in the same company or even an authority in the field where the problem arose, can sell their solutions to the seekers for ten to a hundred thousand dollars. Here's a typical story: a Swiss radiologist, who was thinking about ways to look at bones in 3-D came up with a solution for a petroleum geologist seeking ways to analyze geologic structures in 3-D.[38]

The business world definitely started paying attention to crowdsourcing even closely held data when a failing mining company in Canada, Goldcorp, put all its geologic data—again, usually kept secret—online and offered prizes to anybody who could show where to find more gold. After paying out a half-million dollars in prize money, Goldcorp mined an additional three billion dollars in gold from a previously underperforming mine.[39] Crowdsourced wealth is being created by amateur fashion designers as well as gold mine owners: the one million online Threadless community members upload designs for T-shirts and vote on the best designs, and the weekly winners are manufactured and sold, sometimes bringing significant profits to the designers.[40] Amazon's Mechanical Turk connects people willing to do small piecework tasks for small amounts of money— proofreading documents, classifying photographs, or transcribing recordings—with people and enterprises looking for inexpensive labor—another form of playbor.[41]

Like collective intelligence, crowdsourcing is influencing or transforming many different fields. Ad hoc efforts such as the search for Gray have led to longer-lasting projects to harness mass collaboration in response to humanitarian crises and natural disasters. The rise of "citizen science" is not just about crowdsourcing data gathering and analysis but also portends a time when entire populations of scientific knowledge creators can use their mobile devices as scientific instruments—science as a collective enterprise on a wholly new scale. Businesses are inviting their customers to help design their products. Journalists, activists, and entrepreneurs are exploring "crowdfunding." A group called the First Aid Corps has distributed an iPhone game that enables people to map the location of lifesaving cardiac defibrillators when they find them.[42]

Whether you contribute your computing power or brainpower to a crowdsourced project, or find ways to crowdsource your own tasks, knowing the range of innovations that are popping up in this field can inform your own participation in this mass collaboration skill. You could start out by helping scientists understand the universe or assisting biomedical researchers who are seeking to cure serious diseases. What you learn as a participant could well prepare you to be an instigator as well. One day, you might be able to use your ability to crowdsource projects to cure a disease, create wealth, or fund a creative or humanitarian project.

The flavor of crowdsourcing known as distributed computation was invented because U.S. taxpayers didn't want to fund the search for extraterrestrial intelligence. Radio telescopes suck down signals from outer space and digitize them, and then Search for Extraterrestrial Intelligence (SETI) looks for patterns that might indicate messages from other forms of intelligent life. The amount of data is huge. Lacking public funding for serious computational power, SETI researchers had a brilliant idea. They made available a screen saver (SETI@home) that volunteers could download to their laptop or desktop computer.[43] Whenever a volunteer's computer became idle, instead of going into sleep mode it would download a chunk of SETI's data, start running the data through a pattern-recognition program, and send the results back to SETI@home. As far as anybody knows, no unearthly messages have been detected. At a time when U.S. and Japanese giants IBM and Fujitsu were competing to build the world's most powerful supercomputer, however, a couple million SETI@home volunteers beat the behemoths when they amassed more than twenty teraflops (twenty trillion operations per second) of computing power.

If you want to participate in a more earthly pursuit, you can join Folding@ home and help biomedical researchers crack the puzzle of protein folding.[44] Proteins are the complex molecules that carry out the instructions of the DNA code and enable living processes to function. The way these molecules configure in 3-D space determines the biological activity of the protein. Yet the number of possible ways to fold a protein exceeds the number of atoms in the universe, so a great deal of computing power is needed. When a game version, Foldit, became available on the networked Playstation 3, the amount of aggregated computing power devoted to solving protein-folding problems broke the petaflop barrier—more than a quadrillion operations per second.[45] Today's smart phones are ten thousand times more powerful than the computers used to land humans on the moon, and billions of people are carrying them. What kinds of computing, for commercial and humanitarian purposes, might be possible with supercomputer collectives?

High-powered telescopes have captured images of far more galaxies than all the world's astronomers have time to classify. That's why Galaxy Zoo enlists citizen astronomers. When signing up for the site, volunteers are first shown a short tutorial. They are then shown real images of galaxies, and are prompted by questions such as "Is the galaxy smooth and rounded?" Each galaxy image is shown to ten different volunteers; if 80 percent of the volunteers agree on a classification, it goes into the database. More than 270,000 people have signed up to classify galaxies.[46] In addition to Galaxy Zoo, Zooniverse.org hosts Planet Hunters, the Milky Way Project, Old Weather (recovering weather observations by the UK Royal Navy around time of World War I), Solar Storm Watch, and Moon Zoo. NASA uses volunteer "clickworkers" to count craters on Mars.[47]

Biologists are using Gene Wiki to make sense of the info flood of genetic data now available as the DNA of various organisms has been decoded. Without incentive to share their data, many biologists faced a social dilemma: it was too much work for individuals to contribute their findings about specific genes to collective databases, even though everybody would be far better off if their findings could be aggregated. By swapping out the somewhat-clunky existing database software and putting the data onto an easily editable wiki, the organizers of the Gene Map Annotator and Pathway Profiler were able to create a scientific institution for collective action that made it for easier to map gene expression data.[48] Barend Mons and twenty-three other authors, including Wikipedia cofounder Wales, published a call in *Genome Biology* to annotate a wikified database of information about proteins: "Calling on a Million Minds for Community Annotation in Wiki-Proteins."[49] Encouraged by the success of Foldit, Carnegie Mellon scientists created EteRNA, a game for nonbiologists, allowing them to design new kinds of RNA molecules; each week, the best designs from the gaming community are synthesized and studied in a "wet" lab at Stanford.[50]

I've been interested in Eric Paulos's probes into future media ever since he was at Intel's laboratory in Berkeley, where he enlisted street-sweeping machines to measure air quality.[51] Professor Paulos, now at Carnegie Mellon, believes that the proliferation of inexpensive environmental sensors in billions of future smart phones will make it possible for entire populations to conduct scientific research. Paulos notes that Apple's Nike+iPod Sport kit that includes a digital pedometer with a music player, and the LG Electronics cell phone with a built-in Breathalyzer or blood-glucose sensor point the way.

Our proposal hopes to expand our perceptions of mobile phones as simply a communication tool and to research our envisioned understanding of them as personal measurement instruments capable of sensing our natural environment and empowering collective action through everyday grassroots citizen science across blocks, neighborhoods, cities, and nations. Our near-term goal is build and study a series of mobile devices outfitted with novel sensors along with an infrastructure that provides public sharing and remixing of these personal sensor measurements by experts and non-experts alike.[52]

Emergency services are being crowdsourced by citizens: CrisisWiki, an editable directory of disaster and emergency resources was inspired by HurricaneWiki and the South-East Asia Earthquake and Tsunami Blog.[53, 54, 55] Violence in the aftermath of Kenya's 2007 presidential election stimulated activists to create a platform for crowdsourcing eyewitness reports though SMS or email, and displaying them instantly on a Google map—spawning a whole field of "maptivism." Ushahidi (Swahili for "testimony" or "witness") subsequently made available a platform called Crowdmap to crowdsource crisis information, which was used in Haiti in 2010 in the earthquake's aftermath and again a month later in the Chilean earthquake.[56] After the 2011 earthquake in Christchurch, New Zealand, an instance of a Ushahidi Crowdmap supplied information about the location of food, water, toilets, and medical care. Ushahidi spin-off SwiftRiver, mentioned earlier in regard to the evolution of crap-detection techniques, is trying to combine automatic methods of filtering out inaccurate information with the judgments of networks of trusted humans—a methodology it calls crowdsourcing the filter.

Others are applying similar techniques to the slow-motion disaster of poverty. I remember meeting Nathan Eagle at the Media Lab ten years ago. He was thinking then about how mobile devices could be enlisted to help the economically disadvantaged. Now he has launched txteagle to enable people in poor countries who are out of work to do small, outsourced jobs through their cell phones, such as filling out surveys, translating text, and collecting address data for business directories.[57] The United Nations plans to use txteagle to survey up to a half-million people in seventy countries about their local governance—paying people about one dollar each and reimbursing the cost of text messages for completing the survey. San Francisco–based start-up CrowdFlower has partnered with nonprofit organizations to translate and map text messages from flood victims in Pakistan and earthquake victims in Haiti, compensating workers through PayPal.[58] Can crowdsourcing mobile labor bring globalization's benefits to the bottom of the economic ladder?

Crowdfunding is an emerging variant. Spot.us allows journalists to pitch stories they would like to pursue and enables individuals to pledge financial support; pledges are held in escrow until the journalist's goal is reached. One of my students funded a trip to the Great Pacific Garbage Patch through Spot.us and published her story in the *New York Times*.[59] The crowdfunding platform Kickstarter.com permits anyone to define a project in need of funding, set the rewards (T-shirts, your name in the credits, a signed edition, etc.) for different funding levels, and establish a monetary and time goal. If the funders collectively pledge enough to fund the project within the stated deadline by transferring money to an escrow account, money is sent to the project. When a group set out to create an open-source alternative to Facebook, they sought $10,000—but their campaign brought in $200,642.[60] Filmmaker Franny Armstrong crowdfunded a production about climate change, raising $750,000 from 223 individuals and groups, and then launched the "Indie Screenings model," which allows anyone to buy a license to screen the film, with screeners setting the price.[61] Scott Wilson proposed manufacturing a wristband that would turn an iPod Nano into a wristwatch, asked for $15,000, and raised $941,778 from 13,512 people.[62] Kiva.org matches microbusinesses in the developing world with microlenders, who can help small businesses get off the ground for as little as $25. Inuka.org enables lenders to microfinance projects by women in sub-Saharan Africa. And DonorsChoose.org allows classroom teachers to post requests.[63] Crowdsourcing and crowdfunding are growing so fast that blogs now track daily events.

Sharma at the London School of Economics recently distilled the critical success factors for crowdsourcing projects, which I paraphrase in the following:[64]

1. *Vision and strategy*. This entails the same sense of awe at participating in a worthwhile task that gamers call epic. Participants in crowdsourced efforts must buy into the story of its vision.

2. *Human capital*. It is key to include a large population of people with a variety of skills, abilities, and literacies. In some cases (Galaxy Zoo, for example), some training is required.

3. *Infrastructure*. Without the intangible social and vision elements, open calls for participation will not yield a critical mass of volunteers, but it is the combination of mobile and online tools that make it easy for participants to contribute, and make it easy to aggregate their contributions into meaningful wholes.

4. *Linkages and trust*. Leaderboards (displays names of outstanding contributors), communication with and among participants, and adoption as well as endorsement by thought leaders and populations are essential

lubrication, just as reputation information is the essential lubrication for eBay transactions.

5. *External environment.* The circumstances make a crucial difference in crowdsourced projects, such as whether there is regulatory support or hostility, whether the economic atmosphere is friendly or hostile to entrepreneurship, and if there is are local needs for development or disaster relief.

6. *Motive alignment of the crowd.* This factor involves how much volunteers participate in the vision narrative of the project sponsors and share values with each other.

Crowdsourcing is often centrally controlled, or controlled in a hub-and-spoke manner: many contributions from widespread contributors are centrally collected and aggregated. When voluntary participation is governed in a more decentralized manner and social media are used to organize production, an entirely new form of economic production becomes possible, in addition to the traditional means of the market and firm: social production. Think of it as playbor organized by and for the playborers' benefit.

Social Production

What characterizes the networked information economy is that decentralized individual action—specifically, new and important cooperative and coordinated action carried out through radically distributed, nonmarket mechanisms that do not depend on proprietary strategies—plays a much greater role than it did, or could have, in the industrial information economy. The catalyst for this change is the happenstance of the fabrication technology of computation, and its ripple effects throughout the technologies of communication and storage. The declining price of computation, communication, and storage have, as a practical matter, placed the material means of information and cultural production in the hands of a significant fraction of the world's population—on the order of a billion people around the globe.

—Yochai Benkler, *The Wealth of Networks*, 2006

In 1991, Linus Torvalds, a twenty-two-year-old programmer, posted a message to Usenet, the worldwide online discussion system. Torvalds was trying to create a version of the Unix operating system for use on personal computers and asked if any other programmers wanted to help him. That simple request led to the creation of Linux, a free and open-source operating system that anybody can use without charge and modify, developed by a worldwide community of volunteers. (The free part means both "free as in beer" and "free as in liberty"; the open-source part means that the source code, the human-readable programming instructions that make Linux work, is publicly available for any programmer to inspect and change.) Thousands of programmers who did not know each other (at least at first), located in dozens of countries around the globe, worked outside the organizational structure of the firm as we know it and without financial incentives as we know them to create software that challenged Microsoft's domination.

The methods that made the creation of open-source software possible have led theorists such as Benkler and Steven Weber to talk about social production as a powerful new mode of economic activity. And although Linux software is free, that doesn't mean it is impossible to make money with it. Companies such as Red Hat have made millions by bundling tool sets of Linux software in easily installable form and offering support services for enterprises that want to build specific applications. IBM went from the brink of bankruptcy as a near-monopoly provider of computing equipment and software to the leading open-source services company in the world in just a few years. Social production has proved to be a hard-nosed big business strategy as well as a vehicle for software counterculturists. Don't be quick to characterize it in old-fashioned economic frames like capitalism or socialism, or confine it either to insurgent communities of programmers or technology megacorporations.

Not too many years after Torvalds, in 1994, when I was involved with launching *HotWired*, the "Webzine" version of *Wired* magazine, I came to know Brian Behlendorf, the twenty-three-year-old Webmaster at *HotWired*. He was also organizing programmers around the world to create a free and open-source Web server. Web servers are the software that enable computers to become nodes on the World Wide Web and publish Web pages. Behlendorf, fearing that Microsoft might gain a monopoly in the Web server market, helped gather an informal alliance of volunteers to create the Apache Web server, which is used by about 60 percent of the world's Web sites.[65] Another free and open-source volunteer project, Firefox, is the second most popular Web browser in the world.

In the highly competitive world in which Microsoft came to have near-total control, an operating system that challenged Microsoft Windows, a Web server that contested Windows Web server, and a Web browser that took on Microsoft's Internet Explorer were all created by uncompensated volunteers who contributed their labor and gave their products away for free. Wikipedia content is licensed as free and open source, and only free and open-source tools are used in creating and distributing Wikipedia content. Google and Amazon both employ Linux to run the computers in their massive server farms. Open-source software is by no means a small niche product of a marginal counterculture.

Social production might be a bigger deal than just its impact on the software and network services industries. Ronald Coase won the Nobel Prize in economics for describing how people organize economic production through two different institutions: the market and the firm. In the market, self-interested individuals offer goods and services for sale, and other individuals seek to buy goods and services; buyers and sellers are coordinated through price signals, so decision making is radically decentralized, coordinated by the market itself. In a firm, production of a particular kind of good or service can be done more effectively by employees coordinated and directed by managers. Both firms and markets rely on property, contracts, and laws to enforce these institutions.

Two fundamental assumptions of capitalism and especially industrial production are that the high cost of factories, offices, or other means of production required capital investment beyond what the workers themselves could contribute, and that the management of production as well as profits are controlled by those who provide the capital. When billions of people came into the possession of digital computers and Internet connections, however, a new mode of production began to emerge. In 2002, Benkler, then a professor of law at Yale University, published an influential essay, "Coase's Penguin, or Linux and the Nature of the Firm," in which he proposed that a third means of production, which he called "commons-based peer-production," was coming into being because of three factors. (The penguin is the unofficial symbol for Linux.) First, digital computers are radically inexpensive as well as powerful means of producing certain kinds of goods—among them, knowledge, news, entertainment, education, software, political persuasion, and art—and second, a global network is a radically inexpensive and efficient means of distributing and coordinating labor and knowledge-based products.[66] The means of both production and distribution were no longer limited to capitalists when the workers themselves could own these same means.

One more factor was necessary for social production to work—the institutions for collective action that Ostrom and the evolutionary anthropologists hold so dear. New means of coordinating production, more akin to those that govern virtual communities than either management procedures or market forces, are made possible through the same technologies that are used for production and distribution. Yet the whole system requires a third element in addition to digital media and institutions in order to work: a critical mass of motivated volunteers. Striking at some of their most fundamental assumptions, economists and sociologists must contend with a pivotal question to understand the existence of successful social production: Why would so many people put so much valuable labor into something they are not paid for?

Weber, professor of political sciences at the University of California at Berkeley and director of Berkeley's Institute of International Studies, addressed the key questions of motivation in his book *The Success of Open Source*.[67] Weber cited surveys of open-source programmers who listed, in order of preference, the following reasons for contributing their expertise and labor: the opportunity to learn the programming craft (if you are good, you can gain the attention of those who know more than you; if you err, you will certainly be told about your errors), the pleasure of working on high-quality code, reputation capital (which can be converted into both social and financial capital), contributing to an alternative to proprietary software, and sticking it to Microsoft. I read Weber's book, underlined it, summarized it, and interviewed Weber twice over lunch at the University of California at Berkeley's Free Speech Movement Café about the reasons why people would do something so contrary to the fundamental assumptions of classical economic and collective action theories. Until Linux, the Web, and Wikipedia provided obvious counterexamples, economists and sociologists agreed that people won't contribute without financial compensation to create goods that others can use without contributing themselves.

Weber sketches "eight general principles that capture the essence of what people do in the open source process: Make it interesting and make sure it happens . . . scratch an itch . . . minimize how many times you have to reinvent the wheel . . . solve problems through parallel work processes whenever possible . . . leverage the law of large numbers . . . document what you do . . . release early and release often . . . talk a lot."[68] "Make it interesting" resonates with McGonigal's design principle for epic games: give people a "sense of awe and wonder" along with a lofty goal to work toward. Challenging Microsoft or creating a free encyclopedia for everyone on earth in their native language does make it interesting, and (for some people) evokes awe and wonder.

Open-source communities solve the social dilemma known as "the public goods problem." Where the tragedy of the commons is about overconsumption, public goods dilemmas are about underprovisioning. Why should I pay for public radio when someone else will? This is known as free riding, and the fear that too many others will free ride appears to prevent people from joining the provisioning of public goods. Yet in open-source production, free riding adds value to the product consumed by noncontributors in several ways. First, those who use open-source software even though they don't contribute to creating it are part of growing the installed user base, and the larger the user base, the more power the software has in competition with proprietary programs. Second, even if only one in a hundred users complains about a bug just once in a lifetime of using the software, that information, multiplied manyfold ("leverage the law of large numbers") becomes valuable to the production community, adding value to the product.

"Scratching an itch" is worth unpacking. If I am a programmer, and my company suddenly installed a new printer and no special software existed to address that printer from a Linux system (known as a driver), then I would have to create that driver myself. I might as well contribute that driver to the public code base. That way, when the printer company changes something in the future and I need a modification, there will be a community of users and programmers who can help me. At the same time that I fulfill my own need, I enhance my reputation in the community while also signaling my availability and competence for potential collaborators.

The dominance of economic self-interest and scarcity of economic altruism is so deeply woven into our narrative of how humans operate that counternarratives have been required to explain how social production works. Rishab Aiyer Ghosh analyzed the economic dynamics of open source in his much-cited article on "Cooking Pot Markets: An Economic Model for the Trade in Free Goods and Services on the Internet," comparing open-source software production to a tribal cooking pot into which one person puts a chicken, another adds onions, and each takes out a bowl of stew. Of course, physical stews are what economists call rivalrous goods. If I eat the last spoonful of stew, you don't get any, and the situation is doubly unjust if you were the one who put in the chicken and I am a free rider. Ghosh argues, though, that digital computers and networks make software nonrival: "If a sufficient number of people put in free goods, the cooking pot clones them for everyone so that everyone gets far more value than was put in."[69] The same capability that vexes the music and motion picture industries—the infinite and infinitely cheap power to reproduce entire cultural works, and distribute them widely—is what makes the magical cooking pot possible with digital goods.

Self-election—volunteers choose what to contribute instead of accepting assignments from managers—is important not only as a motivator but also as a way of dramatically lowering coordination costs. By eliminating centralized decision-making structures like managers, commons-based peer production enables innovations to improve the system, repairs to be made to malfunctions, and more efficient systems to evolve simply by providing a structure whereby programmers can jump in where they add value.

According to Weber, four organizational principles are needed for this process, which he labels distributed innovation, to work: "Empower people to experiment. . . . Enable bits of information to find each other. . . . Structure information so it can recombine with other pieces of information. . . . Create a governance system that sustains this process."[70] Weber is a political economist, which means that he looks not only at how resources are allocated and goods are produced but also at the way production processes are governed. The elimination of some management structures by self-election, the channeling of self-interest into public goods by allowing contributors to scratch an itch, and the digital nonrivalrousness of goods produced this way add up to more than just a way to create software. Weber joins Benkler in believing that we are seeing the rare emergence of a new mode of production that could be applied to a wider variety of goods and services.

The early clues provided by Weber, Ghosh, Benkler, and others may enable the social production of important goods in the future by entrepreneurs, following in the footsteps of Torvalds, Wales, and Berners-Lee. If you seek to instigate social production, be humble (whether or not they actually are humble, Torvalds, Wales, and Berners-Lee consistently present themselves in that manner.) Put out an open call, make it easy for people to contribute in as many different ways as possible, supply means for contributors to communicate with each other and technically coordinate their communications, encourage tagging and other ways for information to find similar information or people to find what they need, provide a simple mechanism for adjudicating disputes and deciding what goes into the final product (social production does not have to be wholly democratic as long as it is open to all bona fide contributors; a small set of senior developers, initially selected by Torvalds, makes the final decision about what new code is admitted to the Linux code base, although the best contributions are bubbled up through community consensus), make it easy and inexpensive to contribute self-interested actions to the public good, and make it easy for free riders to help you in some way. For those motivated by reputation capital, provide leaderboards that display the names of frequent and recent contributors.

Commons-based peer production may not be the only kind of systemic change that mass collaboration is bringing to economic systems. How about "collaborative consumption?" In late 2010, a book titled *What's Mine Is Yours: The Rise of Collaborative Consumption*, by Rachel Botsman and Roo Rogers," heralded "the rapid explosion in traditional sharing, bartering, lending, trading, renting, gifting, and swapping reinvented through network technologies on a scale and in ways never before possible."[71] The authors cite marketplaces such as eBay and Craigslist as well as "social lending" sites such as Zopa.com, peer-to-peer travel sites such as Airbnb.com and VRBO. com (a way to search for rooms, apartments, and houses listed for nightly rent by owners), and car-sharing services like Zipcar, Getaround.com, and RelayRides.com. Just as digital media and networks lowered the production and distribution costs of goods such as software and knowledge, peer-to-peer markets lower the cost of trading or sharing goods, skills, and services.

Renting a car has been too much trouble, so many people own private vehicles that remain idle most of the time. What if a rentable car was never more than a couple blocks away, and you could find out where it is, reserve it, pay for it, and then leave it somewhere else when you are done, all via the Web? That's Zipcar's premise—a start-up that experienced a successful initial public offering in 2011. What if you could get a significant discount by joining a number of other people in buying a particular good at a specific time? Start-up Groupon.com, which turned down a four billion dollar acquisition offer from Google, makes Web-enabled group buying easy. Botsman and Rogers see the previous era of consumerism morphing into something that includes a lot more swapping, sharing, and group buying. The four principles that the authors believe make collaborative consumption work mesh nicely with what we know about cooperation theories: trust between strangers, belief in the commons, idling capacity (do you have a room that is not being used or an automobile that is not being driven at the moment?), and critical mass.

Perhaps the most well-known example of social production, Wikipedia, not only illustrates how mass collaboration is changing the way we know and learn but also provides clues to those who would master the collaboration literacies that wiki media make possible.

Wiki Collaboration

When designing software for community, don't make a priori assumptions about how the work needs to be done. Leave it really open-ended, give people the tools they need to talk and give people the tools they need to self-organize, because the procedures change all the time. It's a very human process of people talking and you can't replicate that by baking a centralized voting mechanism into the software. We combine the capabilities of the software with the policies of mediation and arbitration that we've developed, and it's very important that we continually work on maintaining an atmosphere of love and respect. In tech circles it isn't really popular to talk about love, but it's really important—the idea that we're doing something for the good of humanity. We're writing an encyclopedia. It's going to be neutral, it's going to be factual, it's going to be really good quality, it's going to be readable, and we're going to give it to people. If you can let yourself be filled with the love for that shared goal, you can get past a lot of editing differences. Doing something that really matters in the world is one of the reasons we've been able to scale as much as we have.

—Jimmy Wales, Stanford University, 2005

"I'm not all that smart, but I'm very friendly," Wikipedia cofounder Wales told me the third time that he and I were on a panel somewhere in the world. I had learned that whenever Wales travels to deliver keynote speeches or participate on panels, he also spends most of his free time socializing with the local community of Wikipedians at gatherings that resemble parties far more than meetings (and where he is universally known as "Jimbo"). Clearly, he knows something about how to instigate mass collaboration, so I invited him to speak at the Stanford seminar on a New Literacy of Cooperation that I presented in partnership with the Institute for the Future's Andrea Saveri in 2005. Wales arrived at my house in a cab, stayed with me for the night, and was picked up by a Wikipedian after the seminar. He tends to stay with Wikipedians when he travels, and Wales's lifestyle is far from what anybody would consider lavish. He networks a lot. And he gets fired up every time he retells the dramatic narrative of Wikipedia's goals. I was reminded of an aphorism attributed to Antoine de Saint-Exupéry: "If you want to build a ship, don't drum up people together to collect wood and don't assign them tasks and work, but rather teach them to long for the endless immensity of the sea."[72]

We had time during dinner the night before his guest lecture in my Stanford class and over the one-hour drive to Palo Alto to talk about how he had come to do what he did. I wasn't surprised that Wales, like Engelbart and Berners-Lee, was most strongly motivated by the desire to create a powerful tool for all humankind. When he started the project, Wales already had enough money to live comfortably for the rest of his life. He had always been fascinated with encyclopedias, and watched with interest as Torvalds convinced thousands of programmers to help him create a public resource. In 1999, Wales, an early retired options trader, hired Larry Sanger, a graduate philosophy student, to help him develop an online encyclopedia that would be created by volunteers and vetted by experts. They launched a Web site called Nupedia in 2000. Wales and Sanger established a seven-step review process for creating publishable articles. But eighteen months and a quarter-million dollars later, only twenty articles had been published. On January 15, 2001, the founders put their volunteer encyclopedia on a wiki—a Web page that anybody could edit without Nupedia's formal review process. Sanger, who had suggested the name Wikipedia, left the project in 2002. By 2011, ten years later, Wikipedia had more than fifteen million articles in over two hundred languages, all freely licensed for anyone to use, and more than one million registered contributors (registration isn't necessary to contribute; all anybody has to do is click the "edit this page" link at the top of every Wikipedia page). It is consistently one of the top-ten most visited Web sites, and according to the Pew Internet and American Life Project, more than one-third of adult Internet users in the United States consult Wikipedia.[73]

Wikis exemplify the new story about how humans get things done. In 1995, Ward Cunningham created a kind of Web page that could be edited quickly (wiki means "quick" in Hawaiian). In some cases, the page could be edited by an authorized, logged-in member of a specified group. In other cases, Wikipedia foremost among them, wiki editing is radically open—anybody who comes along can click the "edit this page" link. More than one major change in mind-set was detonated by the simple affordance of a Web page that anybody can edit.

The first paradigm shift this young medium provokes is the notion that wikis are documents created by communities rather than authors. The second paradigm shift is to recognize that community is not only key to the voluntary creation of real value (if Wikipedia page views were supported by advertisements, revenue would be in hundreds of millions or billions of dollars); community is the way wikis protect themselves against damage. That protection combines the watchful repair of damage by volunteers with the tool's technical affordances: no draft is ever truly discarded, and it is easy to revert to any previous draft with one click. Whenever a change is made, a new draft is created and the previous draft is stored. The index of all changes, the revision history, is not editable, can be inspected with one click on the "recent changes" link, and visually highlights the differences between any two revisions. Each page on a wiki, in addition to revision history and edit link, has an associated "talk page" or "comment thread," where editors can discuss and argue about the editing process.

You can click around and inspect these aspects of wiki functionality for yourself on any Wikipedia page, and I encourage you to do so. These technical capabilities—easy editing, stored revision history, ease of comparison and reversion, and talk pages—combined with an organization's or community's norms, rules, customs, social contracts, and bylaws, are what make Wikipedia and other successful wiki communities possible. Wikipedia is hardly the only case of paradigm-busting wiki collaboration. By 2007, cell phone manufacturer Nokia estimated that 20 percent of its sixty-eight thousand employees used wikis; by 2006, employees of the Frankfurt-based commercial bank Dresdner Klienwort Wasserstein had created six thousand wiki pages, eliminating much of its email backlog through wiki work.[74]

Does wiki government sound outright crazy? It's already happening. In 2007, New Zealand citizens used a wiki to suggest wording for a new Act of Parliament—an important policing law.[75] The city council of Melbourne, Australia, launched the Future Melbourne wiki, "the city plan that anyone can edit," in 2008.[76] In the United States, New York Law School professor Beth Noveck wrote about "'collaborative governance,' shared processes of responsibility in information-gathering and decision-making that combine the technical expertise of public experts with the legal standards of professional decision-makers. There are plenty of people with expertise to share if their knowledge can successfully be connected to those decision-makers who need it."[77]

I was excited enough by Noveck's ideas when I was investigating theories of cooperation to visit her Manhattan office. Her proposal sounded idealistic and theoretical, but Noveck had a practical and vexing problem in mind—the U.S. Patent and Trademark Office, plagued by a huge backlog of applications. With her "peer-to-patent" notion, Noveck proposed crowdsourcing the vetting of patent applications, using a Wikipedia-like community governance—people with an interest in a particular field can self-elect to offer advice to official patent examiners by annotating and commenting on patent applications.

On June 15, 2007, the Patent and Trademark Office launched an experiment based on Noveck's design, in which the online community not only provided information but also rated the "prior art" (evidence that the patent application duplicated an existing device) submitted by community members, forwarding only the ten best rated to patent examiners. The voluntary contributions from a community of self-elected patent advisers, properly governed and filtered, was meant to serve the patent examiners as a kind of "human database": "By structuring the request for feedback, the agency avoids inviting participation it cannot use. And the public has an opportunity to participate in a way that is directly relevant to and will shape decision-making."[78]

The experiment's results were impressive enough for Noveck to be appointed as the Obama administration's deputy chief technology officer for open government. It's going to take a while for the methods pioneered by Noveck to penetrate U.S. and other government bureaucracies. Wikipe-

dia, however, as a Web-native institution, doesn't have to change old ways of doing things in order to invent new ones.

Understanding why Wikipedia works and how it addresses its vulnerabilities can help anyone who wants to organize wiki collaboration. First, Wales continually reminds the community of its shining ideal—the awe-inspiring epic goal the community shares despite whatever other differences people within it might have. Next, as Nupedia learned, extremely low barriers to contribution and self-election can detonate explosive contribution; the expert on a species of ant and the expert on the history of one block in a small town somewhere each have an easy way to contribute what they know. As in a good game, Wikipedia offers a broad variety of roles to would-be participants, such as readers who point out errors, editors who add and change content, "recent changes" patrollers who revert vandalism, policymakers and administrators, subject area experts, stylists and copy editors, software developers, and proofreaders. Undoubtedly, there are more roles.

Community-based security is another challenge to the traditional narrative of mistrustful creatures who dare not open systems to the entire world because there are some bad actors out there. Instead of employing purely technological defenses or social barriers to entry, Wikipedia makes it easy for people to change a misplaced punctuation mark or delete entire pages. Making it easy to edit makes it easy to damage. Yet one role in the Wikipedia community is "recent changes patrol." Anybody can adopt a page and arrange to receive a notification the moment that page is edited; other tools make it easy to see if the edit is destructive with one glance, and one click restores the prior revision. If more people repair vandalism than commit it, then the system becomes robust even if, paradoxically, it remains vulnerable.

The "Seigenthaler affair" revealed one of these vulnerabilities. A distinguished U.S. journalist and political figure, John Seigenthaler, took Wikipedia to task for a biographical entry that had been altered for a time to imply that he had been a suspect in the Kennedy assassinations—a totally bogus charge.[79] Then there was Dublin university student Shane Fitzgerald, who posted a temptingly repeatable but entirely false quote on the Wikipedia page of composer Maurice Jarre after the French composer's death; before Wikipedians caught and corrected Fitzgerald's misinformation, it was picked up and quoted by several mainstream news outlets.[80] And there were the wikiality and truthiness episodes, mentioned earlier, in which television comedian Colbert urged his viewers to keep editing the

Wikipedia entry on elephants to include the statement that the "elephant population in Africa has tripled over the past six months."[81]

Seigenthaler, Fitzgerald, and Colbert point to glaring vulnerabilities. But those who believe this invalidates Wikipedia's usefulness should also consider the way Wikipedia's social system has built learning and self-improvement into its evolving governance. All these criticisms—and new ones, whenever they arise—can be found in Wikipedia's entry about Wikipedia[82] The community argues about these vulnerabilities and attempts to amend the system in response. When the prestigious science journal *Nature* probed the accuracy of both Wikipedia and Britannica, it found that of eight serious errors in a random sample of forty-two scientific articles, the average Wikipedia entry contained four errors or omissions, while Britannica had three, and each encyclopedia was responsible for four of the eight most serious errors.[83] Community vigilance among Wikipedians extends not only to damaged content on the site but also to bugs in the system, such as those highlighted by the Seigenthaler, Fitzgerald, and Colbert episodes.

Arguing about everything is a core community value in Wikipedia, but it's debating in a specific way. The most powerful and important social norm underlying Wikipedia's social production of encyclopedia pages is what is known as a "neutral point of view." If there are several different opinions about the interpretation of events or data, then each kind of opinion must be stated in a way that adds up to a neutral article. The norm, evoked so often that it is known as NPOV, grew out of the "edit wars" that happened when competing interpretations of an entry collided. You can imagine, for example, what kind of disagreements of interpretation the George W. Bush, Barack Obama, or Israel-Palestine pages stir up. If you are an editor and you have a particular opinion, you are obliged to include the opinions of those who disagree, stated in a manner that those who disagree with you will be disinclined to revert your edit. In fact, the Israel-Palestine page was cited by Wales in his Stanford appearance as a case of two editors—one Palestinian and one Israeli—for whom the word "disagreement" is probably too mild to describe their differences, who have been able to use NPOV to keep the page current as events occur, without sparking endless edit wars.

Wikipedia articles, like most of the Web, differ in quality and veracity, leaving it to the reader to decide how much to trust it. Nevertheless, like most of the Web, the tools are there for anybody who wants to do their own crap detection. Looking at the revisions page frequently reveals who the most prominent editors of a page are, and it's easy to find out how many edits each editor has made on all of Wikipedia as well as how many of that editor's articles have been judged to be outstanding by community vote.

The quality of discussion on an editor's profile page offers clues. And tools such as WikiTrust and Wiki-Watch, described previously, are also available.

Because Wikipedia can be vandalized and is known to contain inaccurate information, most teachers discourage students from using it. In a 2007 speech at Pearson Publishing, boyd, senior researcher at Microsoft Research, made an eloquent case for using Wikipedia in the classroom:

Today's youth have information at their fingertips, but they are constantly being told that this information is inherently flawed and that they should not use it.

Wikipedia certainly has its flaws, but it's not evil. In fact, it's an ideal site for learning how to interpret information. Consider California History Standard 11.1.2 where students are supposed to learn about the cultural dynamics behind the American Revolution. The view from the American and British history textbooks is quite different, yet, the English Wikipedia entry has to resolve these two perspectives. Right now, teachers say that what's in the textbook is right and what's in Wikipedia is wrong. Imagine, instead, if teachers helped students understand why these two differed. Imagine a culture where information is collectively valued, but youth are taught the skills for interpreting it and evaluating it rather than simply being told that everything in the information ecology that they inhabit is "bad" simply because it's not in traditionally vetted sources.

This is a personal pet peeve of mine because if educators would shift their thinking about Wikipedia, so much critical thinking could take place. The key value of Wikipedia is its transparency. You can understand how a page is constructed, who is invested, what their other investments are. You can see when people disagree about content and how, in the discussion, the disagreement was resolved. None of our traditional print media makes such information available. Understanding Wikipedia means knowing how to:

1. Understand the assembly of data and information into publications
2. Interpret knowledge
3. Question purported truths and vet sources
4. Analyze apparent contradictions in facts
5. Productively contribute to the large body of collective knowledge[84]

Wikipedia, though, is not just for information consumers. It's an invitation to participate in the creation of culture. Not everyone can get into an Ivy League university, but everybody has the opportunity to learn how to be a Wikipedian. Contributing to Wikipedia is simple enough. Here's how:

1. Click on the "edit" link at the top of every Wikipedia page and make your edit, adding or subtracting text.
2. Enter a line summarizing your edit. Are you adding a link? Why? Are you adding or removing punctuation? Are you adding or removing content? Why?

3. Save the page.

4. Return to the page an hour and then a day later to see if your edit was reverted, and why. If you disagree, go to the talk page and make your case.

And here are a few words of advice about wiki collaboration in general, gleaned from my own experience and discussions with successful wiki collaborators:

1. *Wikis balance structure and freedom.* A blank page with an overly general mission is daunting. Provide minimal structure at the beginning, a few top-level headings as starting points, with the freedom to branch in many directions. Not enough structure can be as paralyzing to a community as too much.

2. *Provide one or more ways for contributors to communicate with each other.* Give them the freedom to structure their communications the way they want. In addition to talk pages and other administrative discussions on the wiki, Wikipedians also use a chat channel.

3. *Set an epic goal.* It should be something people are willing to volunteer for.

4. *Make as many opportunities for self-election as possible.* Communicate the wide variety of roles that could help. Encourage people to jump in where they can add value. Model these norms.

5. *Set up your decision-making institutions in advance.* Just as software needs a "boot sector" that loads the rest of the instructions, wiki communities need a minimal decision-making procedure at the outset to avoid getting caught up in enthusiasm-draining metadiscussions about governance. Wikipedia started out as a "benevolent monarchy," with Wales making the final decisions in case of deadlock, and although a whole cyberbureaucracy of norms and administrators has grown up, Wales saves the system from death by consensus seeking through retaining (although increasingly rarely using) his veto power. Torvalds fills a similar role in the Linux community. Or if you want a democratic procedure, spell out how to qualify to vote and how voting works *before* you make an open call for volunteers.

6. *Make the recent revisions visible and acknowledge exceptional contributors.* You should also consider a variety of leaderboards, such as, for instance, the most recent as well as largest number of contributions.

7. *Maintain index pages and navigation links.* The overall structure shouldn't grow into a mass of spaghetti.

8. *Fear inactivity more than making mistakes.* If in doubt, just do it; someone can always correct it later.

9. *Provide personalized profile pages for regulars.*

10. *Enlist the entire community in security*. Not every wiki has to be as open as Wikipedia, but if you are putting out an open call, you want to keep your barriers to entry as low as possible. Make it easy to repair damage, and expel and banish bad actors, and then hope that more people with more goodwill than the vandals will heed your call.

11. *To get started, consider a barn raising*. Begin with a face-to-face social event. Ask everybody to bring a laptop and, if necessary, offer instruction. The group can then spend time talking to each other in person while starting to build the wiki.

12. *Encourage emergent leadership*. Explicitly describe and model leadership as community service. You can't push people to grow a wiki; you can only inspire them. Those who can lead by example should be given increased responsibility. Communities shouldn't fear new leaders but rather become enthused about having more leaders, because that means more service. Don't confuse leadership with authority. Minimize individual authority and diffuse it. Everybody should be authorized to act when they see the need.

13. *Set an example of documenting edits*. Encourage editors to fill out a short form summarizing what they did (punctuation, added or deleted material for a reason, etc.). Make everybody's work transparent and visible to everybody else. Encourage each contributor to make it clear to everyone else why they did what they did.

Having taken on the challenges of attention, crap detection, participation, and the five flavors of collaboration, the last of the five new literacies to master is network savvy: knowing how information, power, and social relations flow in a networked world. If attention literacy is the thread that strings together the literacies, network literacy is the fabric of interconnections between human sociality and digital media.

5 Social Has a Shape: Why Networks Matter

One of the key ideas about human social networks is that in the addition of ties between people and specific patterns of ties that obey particular mathematical rules, the whole becomes greater than the sum of its parts. The collection of human beings have properties that do not reside within the individuals, and this collection of human beings is now able to do things that they previously were not able to do. And one of the illustrations or examples that I most like to give about this is something that most people are familiar with from high school or college chemistry and that is the example of carbon. So you can take carbon atoms and you can assemble the carbon atoms into graphite and here we put together a particular hexagonal pattern of ties and you get sheets of graphite and this graphite is soft and dark. Or we can take the same carbon atoms and assemble the bonds between the carbon atoms differently and we get diamond, which is hard and clear. These properties of softness and darkness or hardness and clearness first of all differ dramatically, not because the carbon is different. The carbon is the same in both, but rather because of the ties between the carbon atoms. And second these properties are not properties of the carbon atoms. They're properties of the group, properties of the collection of carbon atoms. Therefore, when we take constituent elements and assemble them to a larger whole, this larger whole can have properties that we could not have foreseen merely by studying the individual elements and properties which do not reside within the individual elements.

—Nicholas Christakis, "The Chemistry of Social Networks," 2010

All the skills you've been learning through this book and your own practice, from infotention to collective intelligence, are deeply intertwined with human and technological networks. We have always lived in a world dominated by networks, from our brain cells to social ties, but we have only recently started to understand how our networked nature affects us. In the late 1990s, scientists began to connect the dots between network structures in physics, biology, sociology, and technical systems, discovering that:

• Networks have structures, and structures influence the way individuals and networks behave.

• Human social networks maintained through the medium of speech go back to the origin of our species. Technologically networked communication media extend and amplify the reach of traditional social networks to make new forms of sociality possible.

• Online networks that support social networks share properties of more general network structure as well as the specific properties of human networks.

A line from a 1990 Broadway play, John Guare's *Six Degrees of Separation*, popularized a notion that turns out to be a linchpin of the new science of networks: "I read somewhere that everybody on this planet is separated by only six other people. Six degrees of separation between us and everyone else on this planet. The President of the United States, a gondolier in Venice, just fill in the names."[1]

In 1967, social psychologist Milgram and his student Jeffrey Travers experimentally discovered this startling idea. Milgram randomly selected 300 individuals in Omaha and Wichita, and gave each of them an information packet along with the name of a contact person in Boston. Each of the initial 300 were instructed to sign a roster included in the information packet and send it directly to the Boston contact if they happened to know them. If they did not know the contact, the subjects were asked to send it to someone they knew on a first-name basis who they suspected might be able to move the letter closer to the Boston contact. Each person in the chain of correspondence signed the enclosed roster before forwarding the packet, thus giving researchers a way to determine the chain's length. Many letters never reached the contact, but 64 of them did. The average "path length" was 5.5. Milgram's results were first published in the popular magazine *Psychology Today*, which made the "small world phenomenon" widely known.[2]

In 1998, Columbia University sociology professor Duncan Watts and Cornell University mathematician Steven Strogatz published a paper in *Nature* about the "collective dynamics of 'small world' networks" that revealed a common underlying structure of all networks that have small average path lengths between nodes despite having large numbers of nodes. Watts and Strogatz also proved that such networks existed in the nervous system of a kind of worm (*C. elegans*) and the power grid of the western United States.[3] Then physicist Albert-Lázló Barabási demonstrated similar structure in the World Wide Web, metabolic networks, scientific collaboration networks, and ecological food webs. Once scientists started looking for it, evidence of small-world networks started showing up in every field—from the spread of epidemics to actors who have been in a film with Kevin Bacon.[4]

In 2001, Watts used email to re-create Milgram's experiment, selecting a random sample of 48,000 senders and 19 targets in 157 countries. Watts found that the average path length was indeed around 6. The *New York Times* reported on a 2010 study by a social media analytics monitoring firm that discovered over 98 percent of the people on Twitter are separated by only 5 steps.[5] In 2007, Jure Leskovec and Eric Horvitz, examining 30 billion instant message conversations among 240 million people, found the average path length among Microsoft Messenger users to be 6.6.[6] Social cyberspaces—whether they emerge from email, blogs, hyperlinks, instant messages, or tweets—are small-world networks, because they are electronic extensions of human social networks.

Usefully for digital citizens, Watts and Strogatz demonstrated how a large network can become a small-world one. You can ascertain the principle for yourself with a pencil and paper. Draw a circle. Around its perimeter, draw dots. Then connect each dot to its two immediate neighbors, one on either side. This is what is known as a highly clustered (dense) network, arranged in a configuration that is known informally as a bucket brigade after the human institution it resembles. Count how many steps it takes to travel from one dot to another on the opposite side of the circle (the path length, à la Milgram). Imagine a circle with seven billion dots on it. Now draw just a few random connections between dots and other dots in other parts of the network, crossing to other parts of the circle instead of restricting the connections to immediate neighbors. It turns out that introducing a relatively small number of random distant links in a highly clustered network transforms it into a small-world network. Do you know someone in Italy? You've now radically reduced the path length between you (and your network) and anyone else in Italy. Networked individuals benefit from having at least a small number of connections to networks that are distant (and different) from their immediate neighbors—and it is even more useful if they are the only person who can bridge two different networks.

We'll see how sociologists are documenting a shift from group-centric societies (in which most of one's friends are likely to know each other) to network-centric societies (in which most of one's network contacts don't know each other). If a network is too densely clustered, Watts and Strogatz discovered, it won't have the short average path lengths that characterize small-world networks in cells or social systems.

If you analyze a small-world network and map the number of connections that each node in the network has with other nodes in the network, you'll find that most nodes will have a small number of connections, while a small number of nodes (supernodes) will each have a large number of connections (a high degree). Your blog probably doesn't get too many inbound links, but Wikipedia gets a lot of links. The Internet is interconnected by a large number of low-degree nodes and a small number of high-degree supernodes, such as Google, Facebook, Wikipedia, or Yahoo! Think of the hub-and-spoke structure of airports versus the structure of the highway system. If you plot out on an x- and y-axis grid the number of nodes against those of inbound links to each node on a graph, you'll get a power

CLUSTERED NETWORK

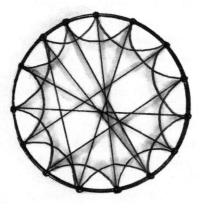

SMALL WORLD NETWORK

law distribution—a curve that's worth knowing about if you are seeking to influence others or do business online.

You already know one kind of statistical distribution curve well—the curve that your teacher graded your class on, with a small number of A and F grades, and larger numbers of B, C, and D grades. Most people know this Gaussian or so-called normal distribution as the bell-shaped curve, due to the shape it takes when graphed. Note that there is a smooth peak at the curve's center, indicating that those who are average in whatever characteristic is being measured are dominant in the population. And note that the tails of the curve on either side of the center's peak are relatively close to the peak. If you are graphing human heights, for example, no humans are as short as six inches or as tall as eighteen feet.

The power law curve looks and acts differently from the normal curve: 20 percent of the power law curve holds about 80 percent of the total area in the graph, which means that the number of nodes of low degree greatly outnumber those nodes with average or greater numbers of connections. These Pareto distributions (named after an economist who noticed its frequent appearance where the distribution of wealth is involved) are sometimes known as "the 80/20 rule," which seems to recur in many kinds of distributions—the holders of wealth in a population, frequency of words in a language, amount of work done by Wikipedia contributors, or popularity of bloggers. It was the occurrence of the power law in the blogosphere that first brought it to the attention of digital culture. Shirky points out that a small number of blogs are the most highly trafficked, and that most blogs have, on average, a small amount of traffic.[7] *Wired* magazine editor Chris Anderson zeros in on the tail of the power law distribution, which unlike the normal distribution, goes off the right end of the page. While there are a small number of nodes in the head of the distribution, there are a much larger number of them in what Anderson calls "the long tail."[8]

A few blogs get a jillion inbound links and hits, and a jillion blogs get a few inbound links and hits. Put this together with the small-world network structure of the Web, and you can see how videos and other Internet memes go viral—some obscure blogger like Bev Harris breaks a story about the Diebold voting machines and others link to it, and then a supernode blog like Andrew Sullivan's links to it. Although the Web affords a large audience to only a few, that large audience is quickly accessible to other publishers when the conditions are right; supernodes diffuse attention to the long tail, the way hub airports feed the regional ones. Anderson is interested in the opportunities made possible by the long tail. While it's more obvious that those nodes in the distribution's head will find ways to profit

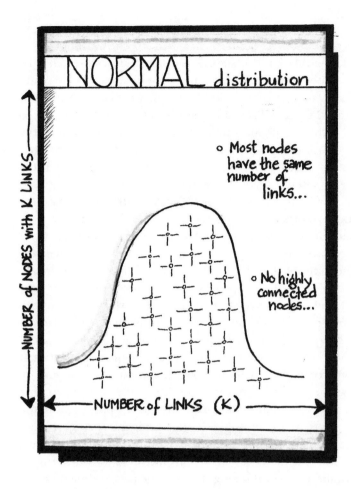

from their popularity, Anderson suggests ways that those unpopular nodes out in the long tail can be valuable as well.

A significant portion of Amazon's income doesn't come from the best sellers but rather from the obscure books and music that aren't easily available in brick-and-mortar stores. Netflix rents out a great number of films that were never hits. If you can aggregate all the fans of an obscure opera singer, people who breed a rare kind of dog, or those who collect antique Balkan tax stamps, you now have a market. The long tail can work for producers as well as consumers. Every niche blogger, curator, and video maker has a connection to potential fans and publics, aided and abetted by search and curation—and every culture producer whose distribution network has a low degree has the possibility of connecting to the head of the power law curve.

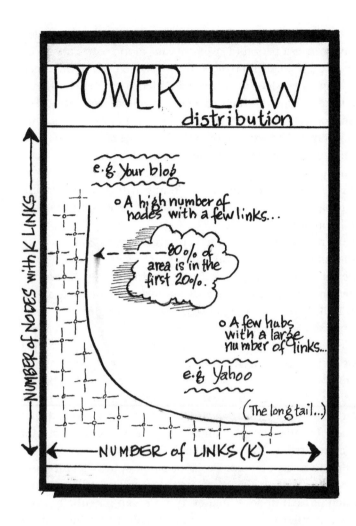

Social network analysis (SNA) of public health data has recently suggested that if your friends' friends (few or none of whom you actually know) are obese, smoke, or are unhappy, you are more likely to be obese, smoke, or be unhappy. University of California at San Diego political scientist James Fowler and Harvard Medical School sociologist and physician Nicholas Christakis reported startling research results in "Dynamic Spread of Happiness in a Large Social Network: Longitudinal Analysis over 20 Years in the Framingham Heart Study," strongly indicating that people's happiness is influenced by how happy their friends, neighbors, and coworkers are.[9] The investigators (who stipulate that happiness is, of course, the product of multiple factors) took advantage of a research database (the

Framingham Heart Study) that happens to contain data about people's health and behavior over a period of twenty years, along with data about their social networks that made it possible to use SNA to explore their relationships. According to Fowler and Christakis, friends of friends' friends have about one-third as much influence as people you know directly. The surprising implication is that at least part of your happiness might depend on people you never met. The research on what has come to be called social contagion also linked obesity, smoking, substance abuse, and other behaviors to social graphs.[10] This research is recent, and conclusions have to be regarded as tentative until others replicate the results with further data, but these findings strongly highlight that network awareness might be vital to your health and happiness.

The amount of freedom each node in a network has to connect with other nodes also influences the network's nature. Answers to the question "Who has the power to communicate with who in this network?" can predict not just the kind of structure but also the relative value that a networked medium is likely to have. David Reed, one of the original architects of the Internet, now an MIT professor, told me about Sarnoff's, Metcalfe's, and Reed's laws ten years ago over lunch across the street from the Media Lab, when I was seeking to understand technologies of cooperation. Sarnoff's law is named after television pioneer David Sarnoff. With a broadcast medium such as television or radio, the value of the network increases *arithmetically* with the number of receivers: add more receivers, add that much more value. Metcalfe's law is named after Robert Metcalfe, creator of the Ethernet, a precursor to the Internet architecture, who declared that the value of a many-to-many network like an Ethernet or Internet increases even more quickly than that of a broadcast network, because adding nodes *multiplies* the reach of each node. When every node can potentially communicate with every other one, then instead of adding another unit of value with each new node, you multiply the number of nodes by itself to determine the number of possible connections. If two nodes are worth four units of meaning or wealth, then three nodes are worth nine, and ten nodes are worth one hundred. The first fax machine was worthless. When there were two fax machines, there was a reason for the owners to have them, but when there were millions of fax machines, the fax network became as valuable as a million times a million.

Reed noticed how those many-to-many networks that also served as platforms for human group formation (such as the Internet and Web) increase in utility radically more rapidly than Metcalfe's law, because the value of each node is multiplied by not only the number of other nodes it can communicate with but also by the potential number of *groups* it can communicate with. Fax machines or telephones don't generally communicate with groups, but humans do. Reed told me that he started thinking about group-forming networks (GFNs) when he wondered why eBay had become so successful: "eBay won because it facilitated the formation of social groups around specific interests. Social groups form around people who want to buy or sell teapots, or antique radios." I quoted my interview with Reed in my 2002 book, *Smart Mobs*: "I saw that the value of a GFN grows even faster—much, much faster—than the networks where Metcalfe's Law holds true," Reed told me, drawing ever-steeper curves on a napkin. "Reed's Law," he continued, "shows that the value of the network grows proportionately not to the square of the users, but *exponentially*."[11]

That means you raise 2 to the power of the number of nodes instead of squaring the number of nodes. Two to the 10th power is about 10 times larger than 10 squared. The value of 2 nodes is 4 under Metcalfe's and Reed's laws, but the value of 10 nodes is 100 (10 to the 2nd power) under Metcalfe's law and 1,024 (2 to the 10th power) under Reed's law—and the differential rates of growth climb the hockey stick curve from there. This explains how social networks, enabled by email and other social communications, drove the growth of the Internet beyond communities of engineers to include every kind of interest group. Reed's law links computer networks with social networks, which are all about group formation and intergroup communication.

In a much-linked article, "That Sneaky Exponential: Beyond Metcalfe's Law to the Power of Community Building," Reed connects these growth laws to the kinds of economic and cultural value each flavor of network tends to create:

There are really at least three categories of value that networks can provide: the linear value of services aimed at individual users, the "square" value from facilitating transactions, and exponential value from facilitating group affiliations. What's important is that the dominant value in a typical network tends to shift from one category to another as the scale of the network increases. Whether the growth is by incremental customer additions, or by transparent interconnection, scale growth tends to support new categories of killer apps, and thus new competitive games.

We can see this scale-driven value shift in the history of the Internet. The earliest usage of the Internet was dominated by its role as a terminal network, allowing many terminals to selectively access a small number of costly timesharing hosts. As the Internet grew, much more of the usage and value of the Internet became focused on pairwise exchanges of email messages, files, etc. following Metcalfe's Law. And as the Internet started to take off in the early '90's, traffic started to be dominated

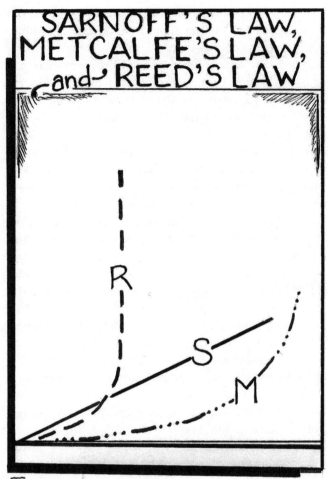

SARNOFF'S LAW,
METCALFE'S LAW,
and REED'S LAW

SARNOFF'S LAW = ————————————

METCALFE'S LAW = — · · — · · · —

REED'S LAW = — — — — — —

by newsgroups, user created mailing lists, special interest websites, etc., following the exponential GFN law. Though the previously dominant functions did not lose value or decline as the scale of the Internet grew, the value and usage of services that scaled by newly dominant scaling laws grew faster. Thus many kinds of transactions and collaboration that had been conducted outside the Internet became absorbed into the growth of the Internet's functions, and these become the new competitive playing field.

What's important in a network changes as the network scale shifts. In a network dominated by linear connectivity value growth, "content is king." That is, in such networks, there is a small number of sources (publishers or makers) of content that every user selects from. The sources compete for users based on the value of their content (published stories, published images, standardized consumer goods). Where Metcalfe's Law dominates, transactions become central. The stuff that is traded in transactions (be it email or voice mail, money, securities, contracted services, or whatnot) are king. And where the GFN law dominates, the central role is filled by jointly constructed value (such as specialized newsgroups, joint responses to RFPs, gossip, etc.)[12]

Many new phenomena, from group-buying services (Groupon.com) to flash mobs, make more sense when viewed through the lens of network structures. As scholar and communications researcher Castells claims, the term "network society" is a much more useful to describe life today than "information society." We've been in an information society at least since Gutenberg. Castells wrote a two-thousand-page trilogy, *The Network Society*, arguing from a comprehensive body of statistical and other evidence that the intersection of social and technical networks is fundamentally reconfiguring human social, political, and economic institutions.[13] In "Why Networks Matter," Castells's introduction to *Network Logic: Who Governs in a Interconnected World?* he lays out seven ways that technologically mediated social networks are transforming society.[14]

First, these networks are global, and the worldwide transit time for information is nearly instantaneous, which Castells contends is the structural basis for globalization. Second, networked organizations outcompete command-and-control bureaucracies. Third, the networking of civil and political institutions is the emergent response to the governance crisis of nation states. Fourth, networks of activists are reconstructing civil society at local and global levels. Fifth, networked individualism, virtual communities, and smart mobs are redefining sociality. Sixth, media space—the public space of our time—now encompasses the whole range of human social practices. Finally, "in this network society, power continues to be the fundamental structuring force of its shape and direction. But power does not reside in institutions, not even in the state or in large corporations. It is located in the networks that structure society."[15] Networks are no longer as simple, rigid, or tightly bounded as power elites have been throughout history. Alternative networks now disrupt and contend with older power structures. Not all these changes are democratic or uniformly beneficial, nor are they wholly predictable, but Castells presents formidable evidence of the ways ubiquitous access to each other and the world's information is reshaping the ways we do everything.

Small worlds, power laws, long tails, Reed's law, network contagion, and network societies are the invisible forces driving many of the social and economic phenomena manifesting today in the behaviors of networked publics. Knowing what these phenomena mean will help you understand the systemic transformation that much of our environment is undergoing. Whether or not you take advantage of them, these network characteristics will continue to influence the way information comes to you as well as how you distribute your own messages, the ways people buy and sell and share, the operations of the levers of power, and the manner in which you and others learn. Moving from the general properties of networks to the specifics of human networks, you'll find that the complex, animated interconnections that make up human social networks have been studied empirically for longer than the Internet has existed.

Fortunately for my purposes here, some of the tools that sociologists developed decades ago are well suited for studying the structure of today's online publics. Early sociologists tended to study the way human groups behaved, but a group isn't the only way human relationships take shape. There are also networks. Instead of examining only the groups people belonged to, some sociologists began asking people to list all the people they interacted with day to day, and then examined the connections between those people. The practice of applying mathematical analysis to these relationship networks to gain useful information about how people behave grew into the SNA discipline—a useful navigation tool for digital citizens.

Social Network Analysis

Billions of people create trillions of connections through social media each day, but few of us consider how each click and key press builds relationships that, in aggregate, form a vast social network. Passionate users of social media tools such as email, blogs, microblogs, and wikis eagerly send personal or public messages, post strongly felt opinions, or contribute to community knowledge to develop partnerships, promote cultural heritage, and advance development. Devoted social networkers create and share digital media and rate or recommend resources to pool their experiences, provide help for neighbors and colleagues, and express their creativity. The results are vast, complex networks of connections that link people to other people, documents, locations, concepts, and other objects. New tools are now available to collect, analyze, visualize, and generate insights from the collections of connections formed from billions of messages, links, posts, edits, uploaded photos and videos, reviews, and recommendations. As social media have emerged as a widespread platform for human interaction, the invisible ties that link each of us to others have become more visible and machine readable. The result is a new opportunity to map social networks in detail and scale never before seen. The complex structures that emerge from webs of social relationships can now be studied with computer programs and graphical maps that leverage the science of social network analysis to capture the shape and key locations within a landscape of ties and links. These maps can guide new journeys through social landscapes that were previously uncharted.

—Derek L. Hansen, Ben Shneiderman, and Marc A. Smith, *Analyzing Social Media Networks with NodeXL*, 2011

Although SNA predates the Web, it has turned out to be a powerful tool for exploring questions about online sociality. The basics are easy enough for anybody to learn, and can be helpful in understanding a surprisingly broad range of things that can happen to us online. It helps to have a pencil and paper at the start.

Most people know how to draw a network diagram. First, draw a number of dots spread around a page in no particular order; next, draw lines between some of the dots. If you think of the dots (known as nodes or vertices) as people and the lines as ways those people could be related (known as ties when talking about the ways in which people are connected or edges when referring to the network's structural characteristics), you now know the fundamental elements of SNA. Ties can represent kinship, friendship, or acquaintanceship, and can also stand for economic transactions, sexual relationships, or prestige hierarchies (think org chart or "above my pay grade"); the kind of relationship that can create a tie is broadly defined. Consider yourself as a node, draw lines between yourself and the people you know and are related to, draw lines between those of your acquaintances who have ties to each other, and what you have is a graph of your personal (sometimes called your egocentric) social network.

You may have heard people refer to Facebook "friends" networks as "social graphs." Search on "visualize Facebook social network" and you'll find a variety of tools for automatically creating a visual diagram of your Facebook social graph.[16] The structures of social graphs and the positions of individuals in the graph can have powerful impact on the nodes (people, including yourself) as well as the graph as a social collective (the social equivalent of graphite or diamond). The business or social success of individuals, whether we survive disease epidemics, our health-related behavior, the reach of messages and transactions, and the effectiveness of political movements, civic organizations, or businesses are all vulnerable to network effects. Rumors, riots, happiness and depression, and knowledge all move through networks with greater or lesser speed.

Social networks can be analyzed in several ways. The strength of ties is one dimension. I'll take that up, but it helps to look first at all the relevant positions of the constituents of social networks. For starters, consider the position of the individual—you—in a network structure. If you list all the people you know and interact with—an "ego network"—then you are in the center and are the only node with some kind of tie to every other person/node. Your mother knows your sister, but your grocer doesn't know your professor; only you know them all. There are many other kinds of human networks besides ego networks. The sexual connections among college students, innovators in corporations, scholars in a particular field, movie collaborators (for instance, the six degrees of Bacon), terrorists, and interlocking directorships of corporations have been subjected to SNA scrutiny. In social systems, the amount of centrality (how well the node interconnects people in different parts of the networks) can be more powerful than the degree (again, the number of ties). In a corporation or a research network, you want to be in a position of high centrality, with many different information vectors coming in and out. In an epidemic, highly central individuals are also more likely to become infected and pass along infections, so occupying a position on the edge of a network becomes advantageous when fear of contagion is important—and centrality could be fatal. Another important SNA term is bridge, denoting a person who is a potential link between two separate networks. Bridge people can benefit from their position, and depending on the bridge person's social skills, so can both networks.

I first learned the jargon of SNA from Marc A. Smith, PhD. In 1991, when I was writing *The Virtual Community*, only a few social scientists were studying behavior in cyberspace. Smith, then a sociology graduate student at the University of California at Los Angeles, was able to answer my question about why people would give time, information, and social support to others online, even if they didn't know the other people well. "Social capital, knowledge capital, and communion," Smith answered—a terse explanation that has stood the test of time.[17] More recently, when I asked Smith to help me figure out how SNA applies to network literacies, he directed me to the work of Mark Granovetter on the ties that connect people in social networks. (My research methodology seems to be: stumble on something, become curious, ask others, and then look where others point. Knowing which others to ask is key to success; indeed, one social network problem in organizations is known as "Who knows who knows what?" Later in this chapter, I'll introduce my power tool for knowing who to ask: the PLN.) I looked at the first place Smith pointed and soon discovered that Granovetter's paper, "The Strength of Weak Ties," in the *American Journal of Sociology*

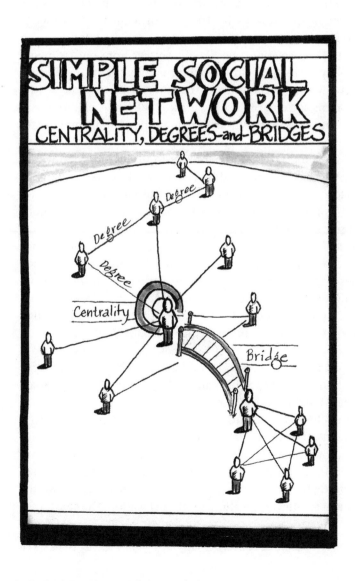

in 1973, was famous among sociologists before and after SNA was applied to life online.[18]

As Granovetter defines it, "The strength of a tie is a (probably linear) combination of the amount of time of the emotional intensity, the intimacy (mutual confiding), and the reciprocal services which characterize the tie."[19] If your house burns down, you are likely to stay with someone with whom you have a strong tie. Yet we all relate to multiple networks of people with whom we have weaker ties. Granovetter notes that the high clustering (density) of strong ties—your strong ties are more likely than your weak ties to know one another—combined with the human tendency to associate with others who share your characteristics and opinion (homiphily)—can limit the amount of information that people can get from our strong-tie networks. Everybody in a highly clustered, homophilous network tends to get the same news, and it is more likely that everybody in a clique (the actual technical term sociologists use) will have the same opinions and access to the same information. In his empirical studies, Granovetter found that numerous weak ties can be important in seeking new information or stimulating innovation. You are more likely to find a job, Granovetter demonstrated, if you have a large and diverse network of weak ties. Note the relationship to small-world networks: highly clustered networks do not exhibit the small-world feature; you need the random distant connections beyond your closest networks.

Granovetter also underscores the existence of absent ties—your friends in high school, former neighborhood, or previous job that you lost touch with. Absent ties have suddenly become more significant in the network society than they were in Granovetter's day. Now that Facebook has changed the historical pattern of leaving old social networks behind when moving to a new school or city, online media make it possible to maintain latent ties at a low cost—which as most Facebook users know, can be both a blessing and a curse. Modern network weavers, as they increasingly call themselves, stress the importance of maintaining a mix of strong, weak, and activated latent (maintenance) ties.

Smith continued his research into the SNA of cyberspaces as a research sociologist at Microsoft Research. Now he is an independent consultant in the application of SNA to digital/social networks. "Be a bridge," Smith advised me, when I asked him for practical tips based on what he knows about the SNA of digital culture. "In social networks, like real estate, the most valuable characteristics are location, location, location," he continued.

Where are you in the network, in relation to others? Which people and which groups connect to each other—or could be connected—through you? If you look at a social graph of a population, the person with the most followers or the most connections might not be the most powerful person in that network. For example, in a network diagram of a population of researchers interested in online sociality, I might identify one person who is the only computer scientist with a strong tie to a sociologist. The sociologists and computer scientists all share an interest in online sociality, but they are two separate kinds of networks with two different kinds of shared knowledge. This person who knows both cybersociologists and social computing scientists can bridge two networks. If that person wasn't there, that lack of connection between networks would be a "structural hole," and the smart person will look for ways to bridge it. Be a bridge. Don't fixate on the number of connections but on the quality of those connections and the diversity of your portfolio of connections. It can be worthwhile to connect to less prominent, less highly linked individuals, if they are different from the other people in your network.[20]

John Hagel and Brown report in the *Economist* that the management of the seven-thousand-person research organization MITRE created internal social media with forums and blogs in order to tackle the question, Who knows who knows what? Hagel and Brown note that for MITRE's management, "The 'Aha' moment was recognizing that their tools enabled some to become 'brokers' between different groups in other parts of the organization."[21] Turner calls such brokers "network entrepreneurs," and sees an emergent profession or position of power resulting from network entrepreneurial talent.[22]

Smith's "be a bridge" advice made such good sense that I asked him for more. "Eigenvector centrality!" he replied, smiling at my baffled expression. "Eigenvector centrality," I shot back, "is a term that is likely to be said aloud only by SNA geeks." He continued: "It turns out that many people outside SNA are familiar with the principle, if not the technical term, because Eigenvector centrality is an ingredient of Google's PageRank." Google and SNA recognize that not every link is equally important. A link from a hub that has many inbound links itself adds to your authority. "So—second piece of advice—get people to link to you. Links to you are proxies for endorsement."[23]

Smith's advice makes sense to any blogger. Most blogs have a form for others to submit suggestions. When I have something newsworthy on my own blog, I submit a link to my blog post to a supernode blog. And most bloggers look at "pingbacks" that tell them who is linking to their posts. Yes, there is structural inequality in the attention economy online, but there are also many routes to wider networks, if you know how to use them. I don't draw supernodes' attention to my own product unless I think those supernode bloggers or tweeters would benefit by passing the link along their publics. Becoming a repeated, reliable source strengthens your tie with a supernode. You have to become a contributing part of the info food chain in order to rise up the chain to leverage supernode attention.

As soon I learned a little bit about SNA, I grew interested in whether it could shed light on perhaps the most crucial question I can pose to my own work: Is life online eroding or enriching our embodied lives? Clearly, some people hang out online way too much. And some people drink or gamble too much. Some people care too much about making money. It's clear that alienation exists, and economic systems and mesmerizing media have something to do with it. The political analyses of media by scholars such as Michel Foucault, Theodor Adorno and Max Horkheimer, Jean Baudrillard, and others, situating deliberately created media illusion within capitalist control mechanisms, are worth keeping in mind. And yet now that we have at least some empirical tools in addition to philosophical analyses, we can ask, In what ways do the new connections afforded by digital networks, both the strong- and weak-tie variety, add more value than they destroy? That inquiry brought me to Wellman, who together with his students at the University of Toronto's NetLab, has conducted extensive, long-term, empirical research into the sociology of life online. I read his papers, and then traveled to Toronto to talk with Wellman and his lab colleagues about our mutual interests.

As a person, Wellman is friendly and soft spoken; the term avuncular is custom made for him. As a scientist who has been pursuing questions about the social impact of media, he is mercilessly empirical, methodologically exacting, and always asking, Where is the data for your assertion that life online is or isn't unhealthy? How was that data gathered and analyzed? After so many years of observing (and, I admit, contributing to) armchair theorizing about what life online means for humanity, I found it refreshing to be offered assertions backed up by scientific observation.

Wellman and I discussed these issues via Skype and Twitter as well as over dinners in San Francisco and Toronto. Like me, he is an active user of social media. Like me, Wellman knows that a critical stance is important for enthusiasts; he's aware that our activities, no matter how pleasurable or empowering, are embedded in systems of economic, political, and social relations. Unlike me, he knows how to apply social science techniques to questions about online sociality. Wellman and his students have published the results of their extensive empirical studies in major peer-reviewed journals. Their findings cast doubt on fears that computer-mediated relationships are alienating people from one another. Most significant, in my opinion, is the way Wellman's NetLab has detailed the social consequences of a shift from a group-centric sociality to what Wellman calls networked individualism.

When I asked Wellman what he thought was the single most important überchange that digital socializing was helping to bring about, his short answer was "the shift from group-centered to network-centered life."[24] A group is densely knit (most members know each other) and tightly bounded (there aren't many connections to people who don't know everyone else), whereas a network is sparsely knit (most members do not know most other members) and loosely bounded (plenty of those small-world-making distant connections to people outside the core). Although groups have been privileged with the warm-and-fuzzy term community, Wellman and others point out that people have always maintained at least a few heterogeneous ties outside their core strong-tie group in order to meet needs that close-knit communities can't provide.

In one fundamental text by Wellman that I assign to my students, he situates present-day fears about cyberalienation in centuries of reaction to modern institutions—and offers evidence that people now find support, information, and a sense of belonging in digital networks as well as physical communities.[25] In the prestigious journal *Science*, describing the conclusion of his team's research into behavior in "Netville," a "wired suburb" of Toronto, Wellman writes,

Computer networks are inherently social networks, linking people, organizations, and knowledge. They are social institutions that should not be studied in isolation but as integrated into everyday lives. The proliferation of computer networks has facilitated a deemphasis on group solidarities at work and in the community and afforded a turn to networked societies that are loosely bounded and sparsely knit. The Internet increases people's social capital, increasing contact with friends and relatives who live nearby and far away. New tools must be developed to help people navigate and find knowledge in complex, fragmented, networked societies.[26]

It becomes tricky to generalize too far in regard to normative ("how it should be for everybody") conclusions. I have personally found abundant support, information, and a sense of belonging—the community attributes that Wellman references—online. What we hold in common is a commitment to examining and reexamining whether we are fooling ourselves, or losing out on something vital through the way we use media.

Nevertheless, despite the importance of individual differences, it's possible to ask some general questions about populations—a fundamental assumption of sociology. When a developer built a new suburb in Toronto in which each household was given the option of access to a broadband Internet connection, Wellman and his team had an opportunity to compare those neighbors who used the high-speed network with those who did not. They found that "those who were part of the high-speed service knew three times as many neighbors as the unwired and visited with 1.6 times as many. Nor was the Internet only used socially: Netville residents used their local discussion list to mobilize against the real estate developer and the local Internet service provider."[27] In this one community, Wellman and his colleagues were able to say with statistical precision that connecting online enhanced most people's off-line lives. It must be said that Canadians affluent enough to move into a high-tech suburb should probably not be used to predict the behavior of people living under different cultural and economic circumstances.

In addition to NetLab, the Pew Internet and American Life Project has conducted extensive scientific polling into the social effects of online media in the United States. In "The Strength of Internet Ties," a report coauthored by the Pew Research Center's Rainie, John Horrigan, Wellman, and Jeffrey Boase, the authors conclude:

Our evidence calls into question fears that social relationships—and community—are fading away in America. Instead of disappearing, people's communities are transforming: The traditional human orientation to neighborhood- and village-based groups is moving towards communities that are oriented around geographically dispersed social networks. People communicate and maneuver in these networks rather than being bound up in one solitary community. Yet people's networks continue to have substantial numbers of relatives and neighbors—the traditional bases of community—as well as friends and workmates.

The internet and email play an important role in maintaining these dispersed social networks. Rather than conflicting with people's community ties, we find that the internet fits seamlessly with in-person and phone encounters. With the help of the internet, people are able to maintain active contact with sizable social networks, even though many of the people in those networks do not live nearby. Moreover, there is media multiplexity: The more that people see each other in person and talk on the phone, the more they use the internet. The connectedness that the internet and other media foster within social networks has real payoffs: People use the internet to seek out others in their networks of contacts when they need help.

Because individuals—rather than households—are separately connected, the internet and the cell phone have transformed communication from house-to-house to person-to-person. This creates a new basis for community that author Barry Wellman has called "networked individualism": Rather than relying on a single community for social capital, individuals often must actively seek out a variety of appropriate people and resources for different situations.[28]

Our lives and societies are networked, but in a paradox that would have made McLuhan smile, network technology has also put the individual at the center, often displacing the traditional role of the place or group. It requires more work on the part of networked individuals to make their way successfully in an always-on, quickly moving world, and a new set of norms and skills—just as the transition from rural, agrarian life to urban, industrial life required. As Turner, Scholz, and Schäfer have pointed out, individual empowerment is constantly in danger of being co-opted and enclosed by commercial interests (for example, playbor). As with the transition to modernity that has taken place over the past three centuries, the transition happening today comes at a cost and yet also has its benefits. What you know, as always, can make the critical difference between being exploited or alienated by your use of social media, and enriching your life and community by your use of the same media.

Networked Individualism

Changes in the nature of computer-mediated communication both reflect and foster the development of networked individualism in networked societies. Internet and mobile phone connectivity is to persons and not to jacked-in telephones that ring in a fixed place for anyone in the room or house to pick up. The developing personalization, wireless portability, and ubiquitous connectivity of the Internet all facilitate networked individualism as the basis of community. Because connections are to people and not to places, the technology affords shifting of work and community ties from linking people-in-places to linking people at any place. Computer-supported communication is everywhere, but it is situated nowhere. It is I-alone that is reachable wherever I am: at a home, hotel, office, highway, or shopping center. The person has become the portal. This shift facilitates personal communities that supply the essentials of community separately to each individual: support, sociability, information, social identities, and a sense of belonging. The person, rather than the household or group, is the primary unit of connectivity.

—Barry Wellman, Anabel Quan-Haase, Jeffrey Boase, Wenhong Chen, Keith Hampton, Isabel Isla de Diaz, and Kakuko Miyata, "The Social Affordances of Networked Individualism," 2003

I remember the exact moment when I recognized the shift to what Wellman's NetLab colleagues label networked individualism. I was in Tokyo, observing the way young people were adopting new modes of communication with cell phones. A local friend of mine, a parent, complained to me: "I don't know my children's friends anymore. They used to have to talk to me when they called my house, but now they just call or text my son's or daughter's mobile phone." In the early years of the cell phone, the phrase most often used at the beginning of a voice or SMS conversation was "Where are you?"—because we had shifted from calling places in the landline era to calling people—who could be anywhere—in the mobile era.

When I started writing about online sociality, the community of interest was the center of cybersociality—the breast cancer chat room on AOL,
the media hangout on WELL, and the newsgroup about your hobby, pet,
disease, or political interest. Individuals looked for online "places" where
they could find others who shared their interests. It was expensive and/or
technical to create an online chat room, BBS, or other social venue. Now
technology has shifted the center from the community to the individual. I
can create a blog, Twitter, or YouTube account in a matter of seconds. I still
frequent online places, but I am also the center of my Facebook and Twitter
networks, I carry my list of contacts on my smart phone, and I check to see
what Yelp users say about restaurants in my neighborhood.

The networked environment, proliferation of networked devices, ease
of summoning our own networks with text messages and tweets, and ways
in which our media powers have shifted our social attention from groups
to networks are a constellation of social transformations that Rainie and
Wellman call "the triple revolution." The drivers of this revolution, according to Rainie and Wellman's forthcoming book, *Networked: The New Social
Operating System*, are the rise of the personal Internet, spread of powerful
mobile information and communication devices, and shift from groups to
networks as the primary focus of sociality.[29]

I've already examined the role of technology and power of participation
in online media. In "The Social Affordances of the Internet for Networked
Individualism," Wellman and his colleagues detail the conclusions of their
research into the social shift toward networks:

Communities and societies have been changing towards networked societies where
boundaries are more permeable, interactions are with diverse others, linkages switch
between multiple networks, and hierarchies are flatter and more recursive. Hence,
many people and organizations communicate with others in ways that ramify across
group boundaries. Rather than relating to one group, they cycle through interactions with a variety of others, at work or in the community. Their work and community networks are diffuse, sparsely knit, with vague, overlapping, social and spatial
boundaries.[30]

My own daily activities today offer a window into how networked individualism operates for many of us. My daughter is traveling in Asia, so I
video Skyped with her this morning—a mediated renewal of a strong tie. I
checked my infotention dashboard for any new information about the topic
of this chapter that might have arrived overnight via RSS. I answered email,
dipped into Twitter (where I asked and answered questions of networks of
people elsewhere in the world I might or might not have communicated

with before), participated in a discussion thread in an online class I'm teaching, used search engines a dozen times, and found an expert by climbing the tree of links from a social bookmarking tag to the first person to post the link to a fundamental article. My wife arrived home and got my attention through an instant message from the other room. She wanted to try a new place for dinner and instant messaged a link to me, so I checked two online rating services to see what my neighbors had to say about the food. I'll get exact directions to the restaurant on my phone on the way. None of these acts were extraordinary. As Rainie and Wellman assert, backing up their claims with substantial empirical data, networked individualism is woven into our lives. The Web is no longer a special place but rather part of most of what we do.

Rainie and Wellman present a detailed description of the kind of people they think will thrive in the emerging environment in which networked individualism plays a strong role. A few short excerpts from their list, along with my annotations, offers a practical resource for those who seek to gain leverage from their knowledge of networked individualism:

- *"Those who can act as autonomous agents to cultivate their personal networks and their 'personal brands':* Social advantages and privileges accrue to those who can 'prospect' for network ties the way effective sales agents can prospect for clients."[31] Smith even has automated this prospecting process.[32] Using his NodeXL SNA software, Smith explores the networks of people on Twitter who participate in a TweetChat using a hashtag that interests him, such as "#socialnetworkanalysis." He looks for both degree and centrality, follows those people, and retweets their most valuable output. Smith has found that 20 to 30 percent of the people he follows this way also follow him back.

- *"Those with bigger and more diverse networks:* Personal networks can now run to thousands of people, if you count the most remote, but still meaningful acquaintances. Although bigger is not always better, those with diverse, broad-ranging networks are often in better social shape and have a greater capacity to solve problems than those who have smaller networks. Those with many functional 'weak ties' can find support and solve problems more adeptly than those who are deeply embedded in a small, tight social network."[33] Remember Dunbar, the anthropologist who believes speech may have evolved from social grooming for the purpose of gossip? His work correlating the size of primate groups with their brain size suggests that the maximum number of people that a human can maintain a strong-tie relationship with is around 150—the somewhat-famous "Dunbar Number."

Further research is required, but common sense indicates that there is only so much time in any day to engage in communications that go deeper than the email, tweet, or text message. The key to a tie's strength is not so much whether your communications are face-to-face or digitally mediated but instead how much of yourself you can put into them. Rainie and Wellman are proposing that networked media make it possible to maintain relationships with larger and more diverse portfolios of weaker ties. This doesn't mean that strong-tie relationships have to go away, as their research confirms. Both strong- and weak-tie relationships can be sustained through media, but strong ties take more time, shared experience, deeper trust, and more frank self-disclosure.

• *"Those who can function effectively in different contexts and 'collapsed contexts':* The act of joining and belonging to multiple groups requires a development of group understanding or knowledge as each has different histories, norms, and folklore. People must learn the ropes in these different milieus. The more gracefully they can do this, the quicker they can assume greater roles within multiple communities and networks."[34] Remember Jenkins's advice (and mine) that the first step in online participation is to understand the social norms of the online context you seek to participate in?

• *"Those who have high levels of trust and social capital:* This is true online as well as offline. A bedrock law of social networking is that people need to discover and interact with those who can provide resources. Humans seem to be hardwired for reciprocity. Social capital has its own rewards as it allows us to gain prestige with individuals or within groups, get things done, and enhance our sense of self. The essential point is that trust and reciprocity are primary currencies for networked individuals."[35]

• *"Those who learn how to manage their boundaries:* As the power of formal, densely-knit groups wanes in light of the buildup of personal networks, personal and community boundaries are less distinct. Does a person want all 300 of her Facebook friends to know what she did last night? With digital technologies, more private information is potentially available to interested members of the public—and to government and organizational surveillance authorities. Networked individuals need to develop new understandings of what to make public, which publics to make information available to, and to intermix technologies of privacy with those of public narrowcasting."[36] We'll see what boyd has to say about this capability, which becomes particularly important with regard to Facebook and Twitter.

• *"Those who like technology and use it enthusiastically and nimbly:* Beyond appreciation of technology and having the skills to use it, media literate people are in better shape as networked individuals in their ability to find information, assess it, react to it, and even remix it with their own spin on it. With this sort of media realism, people can manage their networks better."[37] Note that Rainie and Wellman's "media literate people" also know how to be critical. I know that Rainie and Wellman agree—because I've discussed this with them—that enthusiasm must be tempered with crap detection in order for it to be the kind of media realism they advocate.

- *"Those who manage their time well, especially strategic multi-taskers:* People need to manage their attention more carefully than ever before. Effective networkers exploit this new digital environment more powerfully than those who get lost in their browsing or swamped by information inputs."[38] This is what I call mindful infotention.

Smith along with Rainie and Wellman agree on the importance of social capital—another term originally applied to face-to-face networks that has value to digital citizens as individuals and to the commons.

Social Capital

In all societies, to summarize our argument so far, dilemmas of collective action hamper attempts to cooperate for mutual benefit, whether in politics or in economics. Third-party enforcement is an inadequate solution to this problem. Voluntary cooperation (like rotating credit associations) depends on social capital. Norms of generalized reciprocity and networks of civic engagement encourage social trust and cooperation because they reduce incentives to defect, reduce uncertainty, and provide models for future cooperation. Trust itself is an emergent property of the social system, as much as a personal attribute. Individuals are able to be trusting (and not entirely gullible) because of the social norms and networks within which their actions are embedded.
—Robert Leonardi, Rafaella Y. Nanetti, and Robert Putnam, *Making Democracy Work*, 1993

I learned about online social capital decades before I heard the term, when I was one of dozens in an online community to coalesce into a support network around Philcat, a community member whose son had been diagnosed with leukemia. As I described the incident in *The Virtual Community*, I had come to know Philcat through WELL, an early online community where we both participated in discussions about parenting. We bragged and shared the joy when others bragged. We complained, commiserated, and traded parent lore. We got to know each other. I was an editor at the *Whole Earth Review* at the time, and Philcat was a pretty good freelance writer; I gave him a few assignments. Later, when he was an editor at the *Yoga Journal*, I wrote an article for him. After our online parenting conversations had been going on for months, Philcat organized a picnic for all the virtual communitarians who had engaged in intense discussions with each other daily, but who mostly had not met face-to-face. The picnic attracted a hundred parents and children. We cooked, schmoozed, and played baseball. After that, the Parenting Conference Annual Picnic and Softball Game became a milestone in WELL's cycle of rituals.

Then, late one night, months after the ball game, Philcat started an online discussion in the parenting conference. That afternoon, his teenage son Gabe had been diagnosed with leukemia. His doctors and family weren't awake to share Philcat's midnight fears, but a few of his online friends were available. By the next morning, twenty or thirty people had joined the dialogue, including a medical doctor, nurse, and leukemia survivor. The support community grew as word spread. We didn't know then that online health support communities would become a big deal decades later. In 1989, all we knew was that Philcat could use our help. We passed the hat, raising over fifteen thousand dollars to aid Philcat's family. When Gabe died, the last pews in a packed church were filled with people who knew Philcat from WELL.

Twenty years went by. Philcat and I drifted apart, although we still both lived in the San Francisco Bay Area. Then I was diagnosed with cancer (I'm cancer free now). I started a blog and posted my daily treatment schedule. People I had known from different parts of my life, including dozens of people I knew almost entirely online, began to volunteer to drive me to radiation treatments. One of those people was Philcat. During the drive to and from the treatment facility, I renewed ties with people I had known from one BBS, newsgroup, chat room, Listserve, or another—as well as some of my face-to-face friends and acquaintances. Social capital accrued, was rebuilt, and became convertible into real-world action, initiated and facilitated by ongoing online discussions among people who shared an interest but had not previously known each other.

Another way to look at how online social capital works is the empirically validated value of paying it forward: doing favors for strangers in a network with no anticipation of direct reciprocation. I spend a lot of time answering email from students of virtual community studies all over the world— expecting no direct compensation. When I know the answer to a question I see on Twitter or a blog, I often pause to post what I know. I do it because I consider it my duty to improve the quality of discourse about social media, and I benefit from the efforts of others who feel the same sense of duty. I want to signal to people who take the time to correct misinformation or answer a stranger's question that I appreciate their efforts and pay them forward.

A few years ago, I was invited to participate in a stimulating and lucrative exercise in imagining the future of a major office equipment manufacturer. When I got to the workshop, the person who had hired me said: "Ten years ago, I emailed you to ask how to become a futurist. You gave me a long and detailed reply." I had not been thinking about reciprocation at the time I responded to that graduate student's query, but when I started investigating social capital research, I discovered that answering email from students, posting answers to questions from strangers in online forums, and other somewhat-irrationally time-consuming forms of online participation are practical ways to behave in a network society.

In the aptly titled "It's Not Who You Know, It's How You Know Them: Who Exchanges What with Whom," Gabriele Plickert, Wellman, and Rochelle Côté presented various SNA of relationships in a (physical) Toronto urban neighborhood that led them to conclude: "The evidence is extraordinarily clear on one subject. The overwhelmingly direct cause of reciprocity is giving support in the first place."[39] Or as Wellman put it in a lecture at the Clinton School of Public Service, "The most important criteria for getting help is helping somebody else. If you want help in the future, help somebody now. Pay it forward. We have hard data on that."[40]

Like social networks, social capital is a way to describe an aspect of human behavior that had a rich history long before the Internet came along, but is now an important part of the socializing that online media make possible. Consider the economic conditions of two hypothetical groups of neighboring farmers. Each farmer in both groups has income and expenses alongside an amount of work to get done in order to bring more money in than goes out. One group of farmers doesn't indulge in much socializing or exchange of favors. The other group gathers informally—perhaps its children play on the same ball team or, like the Amish, go to the same church—and exchange favors. For example, if one farmer is sick or injured at harvesttime, the other farmers might take turns helping their neighbor bring in the crops, or the more sociable farmers might lend and borrow tools, increasing the size of each farmer's tool set. The second group of farmers has a sort of wealth that can't be accurately measured by looking at its overall financial income and expenditures. What this group has is social capital—networks of trust and norms of reciprocity that enable the farmers in this group to get things done together that they might not have been able to do otherwise.

We live in societies of laws (the third-party enforcement mentioned in the quote at this chapter's opening), and markets that have written rules and can be measured in dollars, yet we also live in social groups that share human relationships and informal norms. In addition to money, people make use of interpersonal obligations, information exchange, feelings of affection and solidarity, and informal institutions for collective action. Social capital is also key to the power of online social networks, where individuals and groups can cultivate, grow, and benefit from it. The term has a long history, apparently originating around 1916 with studies that sought to explain goodwill, fellowship, and solidarity among people at rural school community centers.[41]

I begin my own framing of social capital with social theorist Pierre Bourdieu's use of the term in the 1970s to describe the resources available to people as a result of durable relationships.[42] In the 1990s, sociologist James Coleman looked at social capital's value to nonelite or marginalized groups as a way of fulfilling needs in the absence of economic capital.[43] In 1990, Wellman and Scot Wortley considered social capital in terms of social ties and social support.[44] (Strong ties are far from the only necessary relationship when it comes to favors, support, and information, as Wellman and Wortley discovered.)

In 2000, Putnam popularized the concern that social capital is also a measure of social cohesion—a measure that seems to be diminishing in the United States, according to the data in his best-selling book *Bowling Alone: The Collapse and Revival of American Community*. Putnam presented a series of demographic and behavioral measures that seem to indicate a steady decline in social capital in the United States in recent decades (fewer people bowling in leagues, for example), which he correlated with the rise of television.[45] A 2011 national survey by the Pew Internet and American Life Project supplied evidence that Internet use may reverse the kind of decline Putnam feared, finding that 75 percent of U.S. adults are active in a group or voluntary association, with 80 percent of Internet users participating in groups compared to only 56 percent of nonusers. Social media users are the most active, with 82 percent of social network service users and 85 percent of Twitter users active in such groups.[46]

In terms of life online, I find it helpful to combine Bourdieu's, Coleman's, Wellman and Wortley's, and Putnam's frameworks. I think of social capital as both an individual's stock of resources that can emerge from sustained social relationships and the capacity of a population—a network or community—to accomplish collective action. An individual can tap into social capital by doing for others and benefiting from what others can do for the individual, but social capital emerges from the interactions of groups and networks in relationship, not from the insulated behavior of any individual. Like the diamond or graphite metaphor for network structures, social capital arises from the shape of ongoing relationships as well as the characteristics of individuals. The two keys to that shape are networks of people who trust each other to some degree, and norms those people share that encourage both reciprocity and occasional uncompensated contribution to a commons.

Although trust and reciprocity seem like fuzzy concepts, there are ways to measure them. Some of the best empirical research on social capital was conducted in the early 1970s, when the government of Italy decided to create a new layer of regional government between the city and national levels. Putnam recognized that this institutional change in Italy constituted a massive social experiment. He mobilized teams of social scientists who measured many aspects of Italian citizens' attitudes and behavior, comparing them with civic and economic statistics. When he wrote about the results of the multidecade research, Putnam posed a number of questions, including, Why is the north of Italy so much more economically successful than the south? His answer, backed by careful long-term surveys and measures, was that historical social circumstances enabled citizens in the north to build social capital that increased both their prosperity and satisfaction with the new civic institutions in comparison with the southern populations. In addition to economic measures, Putnam's team looked at the roots of today's norms and informal institutions in historical social differences.

The city-state-dominated regions in the north had centuries of informal voluntary civic associations such as choral societies, cooperatives, and guilds that led to horizontal social networks. Historically, the south was dominated for centuries longer than the north by the feudal system of the kingdom of Naples. There are fewer horizontal associations in a feudal system, dominated by vertical obligations upward to lords and downward to those lower in the order (such as vassals or peasants). The result, Putnam and his colleagues conclude, was that "some regions of Italy, we discover, are blessed with vibrant networks and norms of civic engagement, while others are cursed with vertically structured politics, a social life of fragmentation and isolation, and a culture of distrust. These differences in civic life turn out to play a key role in explaining institutional success."[47]

Putnam's team asserts that in civic communities, citizens are bound by horizontal relationships of reciprocity, rather than vertical relationships of authority and dependency. Over time, the strands of these relationships weave a kind of fabric that enable the civic community to more easily work around what economists call 'opportunism,' in which shared interests are difficult to realize because individuals in isolation succumb to incentives to defect from collective action. (Ostrom would recognize this social dilemma as the problem of public goods that are underprovisioned because too many people fear that others will free ride on their contributions.) Participation in civic organizations trains people in cooperation skills and strengthens a sense of shared responsibility, thereby building trust that reduces the fear of free riding. These groups don't have to be political; choral societies and soccer clubs knit people together socially and culturally, but the bonds of trust and social networks serve as effective vectors for economic as well as political activities.

The northern Italian cities—Genoa, Pisa, Venice, and later Florence—took off in the eleventh and twelfth centuries in part because the contract and extension of credit were new legal strategies for creating partnerships as well as raising capital. Banking and credit, essential elements for the birth of capitalism, were invented in northern Italy.

Putnam's team doesn't think this was accidental. As Europe arose from feudalism, the bonds of personal dependence (lord and vassal) grew weaker in the northern regions, but in the south of Italy they became stronger. Northern populations learned to be citizens, and southern populations remained subjects. "In the cities, a horizontal arrangement emerged, characterized by cooperation among equals. The guild, confraternity, university, and the commune—a guild of guilds—reflected the new ideals in new institutions."[48] Mutual aid societies flourished in preunification Italy (circa 1850) —pragmatic institutions in which cooperation conveyed benefits on contributing individuals in a changing society. Italian cooperatives grew out of the mutual aid societies. "Networks facilitate flows of information about technological developments, about the creditworthiness of would-be entrepreneurs. . . . Innovation depends on continual informal interaction in cafes and bars and on the street."[49]

Social networks allow trust to spread transitively. People learn that trusting one another pays off, and then institutionalize that learning. Unlike financial capital, trust increases when you use it and becomes depleted if not used. Social capital, unlike conventional capital, is a public good, not the property of any of the individuals who benefit from it, and must often be produced as a by-product of other social activities. Cooperation must also be enforced in a lightweight (and therefore inexpensive) manner by norms, which are less formal, yet frequently as powerful as laws in regulating aspects of social behavior. Because of norms, people in elevators don't usually look at each other; they are engaging in what sociologist Goffman calls civil inattention. People pretend not to see each other's naked bodies in public locker rooms. Norms are generally enforced by fear of both breaking taboos and shaming as informal sanctions (punishments for breaking unwritten rules). "Norms are inculcated by modeling and socialization (including civic education) and by sanctions," writes Putnam and his colleagues.[50] Recall that Ostrom identified graduated sanctions as one of the norms required for successful institutions for collective action.

Observing reciprocity is another norm that is important to the formation of social capital. Reciprocating—paying back—can be specific (quid pro quo) or generalized (diffuse). Diffuse reciprocity means you don't pay back only to individuals but also to the network or community. Communities in which the norm of diffuse reciprocity is high can more efficiently restrain free riding and more easily resolve collective action problems. Networks of civic engagement increase the potential cost to defectors who risk losing benefits from future transactions. The same networks that foster norms of reciprocity also facilitate the flow of reputational information. Remember Dunbar's theory about speech emerging from social grooming in order to facilitate gossip—the dissemination of reputational information? When someone in your network asks for a favor, and it's easy to grant it, do it even if you don't know the person. The word might get out that you grant favors and therefore are a good person to do a favor for. Passing along information about untrustworthy actors is tricky business; if you are going to do it, you should crap detect your sources thoroughly. Try trusting strangers in small ways if they are in one of your networks. Do pass along information about trustworthy actors.

None of this is rocket science. Indeed, Wales told me that most people learned on the playground most of what they need to know to be good Wikipedians. Tune your networks; pay attention to who you grant some of your attention. Feed your networks; give freely what you know people can use. (I learned "feed your networks" from Kevin Kelly's New Rules for a New Economy in 1996.)[51] Increase trust by avoiding bad actors and spreading the word about good ones. Recall how collective intelligence research suggests that social intelligence (rather than strictly intellectual capability) increases the network's intelligence—and that increasing the number of women can increase a network's social intelligence. Networks that are more diverse, in which individuals are different from each another and connect with different networks, provide richer environments for circulating knowledge as well as social capital. A certain amount of clustering is good. People need to know what shared interests or qualities connect them to one another, and if trust is at least partially transitive, then I can take the word of a person

I trust highly that a third party is trustworthy. Too much clustering, however, creates social and knowledge insularity.

Just as there are different strengths of ties (that is, strong, weak, and latent), social capital ties come in several different flavors. *Bonding* social capital refers to ties between people who share strong mutual contexts and invest relatively heavily in their relationship, such as strong-tie friendships, family, neighbors, and coworkers. *Bridging* social capital is a function of weaker and more distant ties—again, between people who have more in common than not. *Linking* social capital involves ties to people in dissimilar circumstances and communities, or the kind of ties that are necessary for small-world networks. Bonding capital increases feelings of solidarity, trust, and specific reciprocity. Bridging social capital helps cliques to break out of their insular worldviews and bring in external information, and assists in diffusing information across multiple networks. "Bonding social capital consists of a kind of sociological superglue, whereas bridging social capital provides a sociological WD-40."[52]

A recent study examined the relationship between Facebook use and the formation and maintenance of social capital, surveying college students regarding their relationships and comparing the survey to their Facebook behavior. It found that the increased use of Facebook among this population strongly increased all three types of social capital, with the strongest relationship to bridging social capital. In addition, the Facebook researchers identified a dimension of social capital that online networks magnify in ways that traditional face-to-face ones do not: "one's ability to stay connected with members of a previously inhabited community, which we call maintained social capital."[53] Facebook, as most people know, is not an unalloyed social benefit. Privacy and reputation concerns can balance or outweigh social capital benefits, as I'll discuss shortly. But this study offers empirical evidence that knowing how to use Facebook can increase your ability to cultivate and harvest social capital.

Recall Smith's advice about "be a bridge," and the potential power in being able to fill "structural holes." Ronald S. Burt, a University of Chicago sociologist who is also an executive at Raytheon, a large electronics manufacturer and defense contractor, investigated the importance of structural holes to organizational innovation. Using archival and survey data along with SNA techniques, Burt studied the people who managed the worldwide supply chain for Raytheon, and discovered that those who actively bridged structural holes gained a competitive advantage for themselves, their division, and their company by delivering more good ideas (Burt's paper on this study is titled "Social Origins of Good Ideas"). He asked several hundred managers to write down their ideas to improve Raytheon's supply chain management, and then asked other executives to rate the ideas, finding that managers who discussed ideas beyond their work groups consistently made the highest-rated suggestions.[54]

Burt also found four different levels in the organization through which people could create value by being brokers between networks: They could "make people on both sides of a structural hole aware of interests and difficulties in the other group."[55] People who transfer best practices from one network to another perform a second, "higher level of brokerage" (say, "the tribe in the next valley invented the wheel, so we don't have to reinvent it"). A third level of brokerage involves making networks that were apparently irrelevant to each other aware of their potential commonalities. Combining and synthesizing elements from disparate networks is a fourth level of brokerage. Intensification of these kinds of brokerage may be one of the most essential cultural values of cities. Another study of innovation at a broader scale than Burt's, conducted by computational social scientist Samuel Arbesman at Cornell, has provided SNA evidence that larger cities generate more connections between different kinds of people and networks, leading to higher levels of innovation.[56] Add to "it's not what you know, it's who you know" the addendum that success also depends on "how different the people you know are from each other."

Turner has devoted considerable scholarly resources to tracing some of the roots of digital culture back to the *Whole Earth Catalog*'s precomputer counterculture. In his book *From Counterculture to Cyberculture: Stewart Brand, the Whole Earth Network, and the Rise of Digital Utopianism*, Turner focuses on Stewart Brand (and me) as examples of an emerging social role of network entrepreneurs. Brand's *Whole Earth Catalog*, Turner claims, was not just a kind of independent living catalog for off-the-grid (or fantasizing about doing so) counterculturalists but instead the intersection of diverse social networks that Brand brought together—the people who would come to be called environmentalists, those interested in tools and technologies including the new personal computers, the independent living people, and the alternative energy and holistic health care people.[57]

I interviewed Turner about network entrepreneurship in the context of this book, asking him what might be useful for readers to know to succeed online. "I ended up coming up with this term, network forum, to try to describe the places, physical or textual, online or off-line, where different networks are brought together by an entrepreneur," Turner told me.

An entrepreneur may recognize different social worlds and might be a kind of peripheral member of different social worlds, but unless they have a place to bring those people together, those worlds never actually meet. When they meet, they need to not only come together in some place. They need to do something together. So Stewart Brand would have these wonderful festivals to support them like Alloy in 1959, where he brought together technologists, counterculturists, and other kinds of folks to build a camp for a weekend. Burning Man would be another example of this. Everybody gets together. They make art. Half of the fun of the gathering is seeing all the different tribes that are represented there. So a network forum to be successful needs a defined space. That can be a real physical space, or it can be an online space, or a textual space like a catalog or magazine. It needs members from different social intellectual communities and it needs a gathering kind of host person—a person who pulls people together and helps them do what they do better.

Another way to think of the host person, the entrepreneur, is as a "Barnum." A Barnum is somebody who organizes a circus, even though they themselves may not be able to do any of the special things in the circus. P. T. Barnum couldn't fly a trapeze, couldn't ride an elephant, couldn't ride a horse bareback, and yet he became arguably the biggest, most visible voice for the circus of the century. So what did P. T. Barnum do? Well, he built rings and spotted performers and invited the performers into the rings. Then he turned around and said, "Ladies and gentlemen, we have a circus, and I'm your ringmaster." He gave everyone a great and entertaining show, but he also gave himself a lot of authority with regard to the action that is going on in the ring. He became the ringmaster. And I think that's the work of the network entrepreneur. The rings and the tents, so to speak, are the network forum.[58]

I asked Turner if he could think of examples that weren't countercultural. "Online neighborhood groups are actually especially good for this," he answered.

I recently bought a new house. I'm living in a new neighborhood. And I'm discovering that there's an online neighborhood chat. We all come from different worlds. Some of us work in the tech industry. Some of us are retired. Some of us have new kids. Some of us are students. It's a diverse community. Online we find small things that we care about. For example, sidewalks, flowers, the state of gardens. We talk about those things, but as we do that we also learn a lot about the world that we come from. So I'm learning quite a bit about Google from the neighbors who work there. I'm learning quite a bit about social services for retired members. The chat makes me a better citizen and a better citizen of my neighborhood.[59]

In fact, sociologist Keith N. Hampton and his colleagues Chul-joo Lee and Eun Ja Her have produced extensive data, reported in "How New Media Affords Network Diversity: Direct and Mediated Access to Social Capital through Participation in Local Settings," so Turner's example appears to be more than anecdotal.[60] Turner added:

I think the whole idea of the network entrepreneur in the network forum scales down as well as up. Stewart Brand did it on a big level. But you really can do it with your friends. I think the biggest challenge to effective network entrepreneurship is being open to people who are very different from yourself and inviting them in. I think that's the biggest challenge. You have to not just live in your home world but be open to members of other worlds and open to projects that might bring people from other worlds together with yours.[61]

Thinking about your social relationships in terms of maximizing your social capital can be useful as well as beneficial to you and others, just as striving to be mindful of how you deploy your attention, approaching online information with an investigative crap detector, knowing how to be an online participant and collaborator, and knowing how to persuade others to collaborate can be useful as well as beneficial. But humans are not reducible to strictly economic terms, and several social scientists warn that the use of the term capital can be dangerously misleading given that not all social relationships are strictly utilitarian. Neither is the phenomenon called social capital limited to the prosperous; the ability to get things done with friends, neighbors, and networks is vital to those at the bottom of the economic ladder.

Social capital is a useful tool, but should not be the only one in your tool set. If all you use is a magnifying glass, all you see are expanded versions of small things. If all you use is a telescope, all you see are unrealistically close-up versions of faraway phenomena. Empathy, friendship, and community always have heart and soul if they are to be authentic. Too much calculation hardens hearts and deadens souls. Video Skype with your daughter when she's faraway, but put down the smart phone and look her in the eye when she's in the same room. Join the neighborhood chat *and* talk over the garden fence. Think of the tips I've gleaned about maximizing social capital as lenses for seeing your social networks more mindfully and productively. At the same time, always keep in mind that you are never seeing the whole or living through just online communication channels, and you can never truly characterize families and communities only by their network benefits.

A good example of a network that can (and should) be cultivated and authentic, instrumental and sociable, is the PLN—something all digital citizens need to learn to grow and maintain.

Tuning and Feeding Personal Learning Networks

If individualized learning is chained to a social vision prompted by "prisoner dilemma" rationality, in which one cooperates only if it maximizes narrow self-interest, networked learning is committed to a vision of the social—stressing cooperation, interactivity, mutuality, and social engagement for their own sakes and for the powerful productivity to which it more often than not leads. The power of ten working interactively almost invariably outstrip[s] the power of one looking to beat out the other nine.

—Cathy N. Davidson and David Theo Goldberg, *The Future of Thinking*, 2010

When I started using social media in the classroom, I knew something about the technology, but less about the pedagogy that it could support. I looked for and began to learn from more experienced educators. I searched on terms such as "educational technology," "social media pedagogy," and "wikis in the classroom," noting names of authors while I collected links, opened tabs, tagged bookmarks, and added radars to my dashboard. I looked at the people who bookmarked a large number of sites with tags like "edu-tech" and "educational technology," inspected their other tags, and noticed who were the first to bookmark the most popular sites. I harvested the names that recurred, found their blogs and Twitter user names, and then started reading them. I followed their blogrolls and added more experts to my list. I tried to comment usefully on their blog posts and tweets.

When I began to understand who knew what about social media in education, I narrowed my focus to the most knowledgeable among them. I paid attention to the people who the savviest social media educators paid attention to. I added and subtracted voices from my attention network, listened and followed, and then commented and opened conversations. When I found something I thought would interest the friends and strangers I was learning from, I passed it along through my blogs and Twitter stream. I retweeted (broadcast another person's tweet to my followers, together with the originator's Twitter name). I asked questions, asked for help, and started providing answers and assistance to those who seemed to know even less than I did.

The educators I had been learning from had a name for what I was doing: growing a PLN. So I started looking for and learning from people who talked about how to cultivate a PLN. One of the real masters of the PLN craft is Shelley Terrell, an educator of educators I discovered in my first search for social media classroom adepts. I ended up blogging about her for the MacArthur Foundation's Digital Media and Learning Web site, DMLcentral.net.[62]

Terrell taught me a simple exercise for getting across the power of PLNs in a classroom. I asked people to ask questions of each other, one to one, about things they would like to know regarding our course's subject. "What's the difference between a blog and a wiki?" for example, or "What do the letters RSS stand for?" Whenever they couldn't come up with an

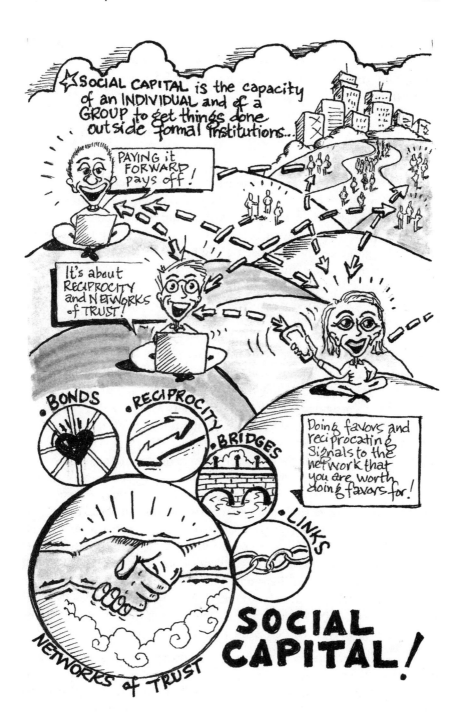

answer between them, I asked the pairs to write down the questions that stumped them both. Then convening the class as a whole, a few of the pairs asked aloud the questions they couldn't address. Almost always, someone else in the class knew the answer. Apply the same principle to online networks that can contain hundreds of people all around the world, personally selected by you because of their expertise in the topics of interest, and you have a PLN.

I learned from my PLN that a PLN is at the same time my personally curated network of people I want to learn from and a network that learns together. It wasn't too far a leap from there to the notion of learning community. And that's when my students and I turned into colearners. Like online social networks and social capital, PLNs are a Web-amplified version of something that people have been doing for a long time. People who are drawn to independent learning look for other students, teachers, and information sources. When search and the Web enable us to find the experts and teachers we seek, and when others can find us to learn from too, the practices of independent learners in the era of paper texts are morphing and multiplying through online social networks and knowledge repositories. PLNs connect the colearners with the texts, videos, open-courseware lectures, info radars, and social media discussion platforms available online—and PLNs teach themselves how to better use PLNs.

One night, engaging with my PLN and others who paid attention to me by following my Twitter, I started musing about how to create PLNs. What should I tell my colearners about my experience? In a series of 140-character tweets, I broke the PLN-cultivation process down into eight processes: *explore, search, follow, tune, feed, engage, inquire,* and *respond*. In between my own tweets, others who were on Twitter at that time began to respond to me, agreeing or suggesting tips. Others used curation tools to collect my PLN how-to tweets into a single addressable page, which I copied to my own teaching notes blog. I then retweeted the links to my curated tweets to my network of Twitter followers. Bits of information and discussions of what they mean move in loops and networks, links and collections, and in the process, are transmuted into knowledge. Here is the advice I gave in that short episode on Twitter, augmented by some of the recommendations I received by others in my PLN while I was emitting these bursts of lore:

Explore multiple media—blogs, Twitter, Facebook, social bookmarking sites, question-and-answer sites, and people you meet face-to-face. Keep tabs on what you find, but your objective is to explore the space of your interests. PLN candidates will emerge only after sufficient exploration. While you explore, you will meet others. Maintain openness to serendipitous encounter.

Search after you have explored enough to get some sense of the field, community, discipline, and subculture. Use the terms you've discovered to search the Web, blogs, and Twitter for experts. Twitter lists and Twitter list compendiums like Listorious are attention-effective ways to find candidates. If you are looking for scholarly expertise, use Google Scholar and Harzing's Publish or Perish.[63, 64]

Follow candidates' activity streams through RSS and Twitter, YouTube, Quora, Tumblr, Posterous, Scoop.it, Diigo, Flickr, and so on, if applicable. Ask yourself each time you look at their output why you added them, and whether their posts or tweets, photos, bookmarks, or videos have been worth your attention.

Tune your network by dropping people who don't seem worth spending attention on regularly. Reciprocity is *not* expected in following activity streams. Follow people *only* if paying attention to them increases your knowledge, or inspires or amuses you. Add new candidates frequently. When you value someone's output highly, begin looking at who they pay attention to. Test and prune. Add, observe, keep, or delete. Turn what you've learned about microdecisions and infotention on your PLN.

Feed the people who follow you by sharing value when you find or create it, whether it is informational, social, or entertainment value. Feed the people you follow by directing specific pieces of information their way when you know they will find them valuable.

Engage the people you follow. Be polite as well as careful about making demands on their attention. Retweet them when you think their contributions can provide value to the people who are spending their own attention on you. Comment on their blog or reply to their tweet when you have something helpful, informative, or (this is tricky with people you don't know personally) entertaining to say.

Inquire of the people you follow and those who follow you. Ask engaging questions. It's always a bonus if the answer can be useful to others in your network. A well-tuned, well-fed PLN can be astonishingly, magically, precisely, and promptly useful. But don't ask anything you could find out with a little googling or two minutes on Wikipedia.

Respond to inquiries made to you. It's only polite—which is not only nice but also a signal to others that you are a contributor. Contribute to diffuse reciprocity. Feed the network if you know what someone needs to know, even if you aren't directly reciprocating a similar favor from that person (or expecting direct reciprocation from them).

Hang out, as Mizuko Ito advises, on Twitter, a chat channel, a message board, a YouTube channel, or a community of commentors on a popular blog. You'll learn things and meet people. Mess around by trying experiments with others and collaborating for the fun of it. Eventually you will geek out by inviting your PLN to help compose advice on cultivating a PLN.

I can't leave the subject of network awareness without dealing with the online social network that has more than a half-billion members globally: Facebook. Important issues of identity, community, friendship, privacy, reputation, and surveillance have arisen because of this site, and are likely to continue. Knowing the facts of Facebook life is, for a growing portion of the world's population, a twenty-first-century survival skill.

Facing Facebook's Facts of Life

Imagine how creepy it would be to wander into a co-worker's cubicle and discover the wall covered with tiny photos of everyone in the office, ranked by "friend" and "foe," with the top eight friends elevated to a small shrine decorated with Post-It roses and hearts. And yet, there's an undeniable attraction to corralling all your friends and friendly acquaintances, charting them and their relationship to you. Maybe it's evolutionary, some quirk of the neocortex dating from our evolution into social animals who gained advantage by dividing up the work of survival but acquired the tricky job of watching all the other monkeys so as to be sure that everyone was pulling their weight and not napping in the treetops instead of watching for predators, emerging only to eat the fruit the rest of us have foraged.

Keeping track of our social relationships is a serious piece of work that runs a heavy cognitive load. It's natural to seek out some neural prosthesis for assistance in this chore. My fiancee once proposed a "social scheduling" application that would watch your phone and email and IM [instant message] to figure out who your pals were and give you a little alert if too much time passed without your reaching out to say hello and keep the coals of your relationship aglow. By the time you've reached your forties, chances are you're out-of-touch with more friends than you're in-touch with: Old summer-camp chums, high-school mates, ex-spouses and their families, former co-workers, college roomies, dot-com veterans. . . . Getting all those people back into your life is a full-time job and then some.

You'd think that Facebook would be the perfect tool for handling all this. It's not. For every long-lost chum who reaches out to me on Facebook, there's a guy who beat me up on a weekly basis through the whole seventh grade but now wants to be my buddy; or the crazy person who was fun in college but is now kind of sad; or the creepy ex-co-worker who I'd cross the street to avoid but who now wants to know, "Am I your friend?" yes or no, this instant, please.

—Cory Doctorow, "How Your Creepy Ex-Co-Workers Will Kill Facebook," 2007

If you know what you're doing with Facebook, you can increase your social capital and that of your friends. You can build a digital portfolio that will help you obtain jobs and promotions. You can market your start-up or cause. You can plot revolution. If you don't know what you are doing with Facebook, you will stream private information about your behavior to corporate, law enforcement and intelligence agencies, credit bureaus, your ex-spouse, your nosy neighbors, or the creepy coworker, "the crazy person who was fun in college but is now kind of sad."[65] You will lose that job or promotion. You'll have a lot of explaining to do to extricate yourself from socially awkward positions. The secret police will know more about your revolution than you do. And the hard part is that there can't be an infallible, up-to-date guide to navigating the pitfalls of Facebook, because Facebook keeps changing its privacy policies along with the ways in which it enables or forces its members to interact with each other.

For example, as the *New York Times* mentioned on the day that this paragraph was first drafted, Facebook changed its users' privacy settings to allow a feature that uses facial recognition algorithms to automatically detect a user's face in an image, and then alerts the user's Facebook friends to tag them.[66] The day after the new feature was reported, Facebook admitted that it had screwed up . . . again.[67] You have to know where to go on Facebook's increasingly complex control panel to turn off this feature. You can't opt out of being tagged in a photo without your consent or knowledge; you can only "detag" photos of yourself after you become aware of them. This was only the latest in a long series of privacy gaffes going back to the dawn of Facebook's history. (In August 2011, Facebook changed their privacy settings again, perhaps in response to the rollout of Google's "Google+" social network service. These new settings gave Facebook users greater control over who they share their updates with.)

When Facebook first rolled out its "News Feed," millions of unwarned users reacted angrily to the news that their status updates and other activities would be broadcast instantly to their social graph.[68] In the olden days before News Feed (2006), "there was a huge backlash from users who called the site 'Stalkerbook,' and protested that it was creepy to constantly see what their friends are doing."[69] A student at Northwestern University created a Facebook group to protest the News Feed; the group grew to 284,000 by the next day, and the day after that, Facebook founder Mark Zuckerberg was forced to publish an apologetic response.[70]

Five years later, the News Feed is a fact of Facebook life, and a new generation of Facebookers consider it normal to see what their friends are doing at all times, and know where they are doing it. In 2007, Facebook began broadcasting members' purchases—the infamous "Facebook Beacon" campaign, a feature that was withdrawn after a sufficient number of people let Facebook know that they didn't want the world to know that they had purchased a surprise diamond bracelet for their spouse, or someone who was not their spouse.[71] If you want to spend an instructive hour, search on "Facebook privacy controversy" or "Facebook privacy gaffe." In fact, you should create an RSS feed from Google, Bing, and Yahoo! news searches on those phrases if you plan to continue being a Facebooker.

I begin a discussion of Facebook literacy with privacy concerns because of these issues' power to affect our lives. But Facebook is also causing us to face a redefinition of what friend means and makes it increasingly difficult to behave differently in front of different publics (the different faces students show their classmates and parents is the classic, but far from only example of what boyd calls "collapsed contexts"[72]). Through its massive data mining and microtargeting of advertising, Facebook also takes the notion of playbor to the point where it threatens to enclose the open Internet into a commercially driven web of monetized "likes."

Fortunately, the skills you've already started to learn will come in handy in a Facebook-colonized world. Your practice of mindfulness regarding your social media use, for one, should focus your awareness on the personal costs of maintaining or not maintaining a Facebook profile at all. Deleting or not creating a Facebook profile in the first place is definitely a legitimate choice, and listing the pluses and minuses of Facebooking is an exercise I recommend to everyone, whether you are Facebook skeptic or enthusiast.

I offer advice here on how to use Facebook effectively, but do not intend to persuade anybody that Facebook is essential. I use Facebook and allow it to use me—up to certain limits—because it is more useful to me than not. But one day, I might delete my account—or however much of it Facebook will allow me to delete. Crap detection and network awareness, in particular, can be useful in Facebookville. Knowing the subtleties of writing a profile, etiquettes of friending along with accepting or refusing friend requests (a convention has emerged of using Friend to distinguish the Facebook variety from the traditional meaning), how and why to comment, how to use your status updates, and how to track others' News Feeds is vital lore not only for young people who are just starting to leave digital footprints but also for their parents and any newcomers who begin making online statements that are probably more permanent, searchable, and revealing than they suspect.

When you compose your biography, think about whether you want the world, everybody in your friends' networks, or just your friends to see it, and then use Facebook's privacy settings to consciously control your boundaries to the degree Facebook allows. Think twice about trying to convey humor or irony in your profile picture, since these intended nuances are notoriously open to misinterpretation online. When considering a friend request, ask yourself what boundaries you want to set for this person. Whether you are creating your first profile or have had a Facebook account for some time, you will benefit from looking for the "Account" link, and then carefully going through the submenu for "Privacy Preferences"—and the submenus under that. Look at who can see your biography, the links and photos you share, your status updates, and your birth date, and ask yourself whether you want to restrict that information to publics you can designate. Scrutinize your profile through the eyes of an identity thief. You might be comfortable letting your friends know your whereabouts, phone number, and mother's maiden name, but do you want your friend's friend's crack dealer to know that much about you too?

While you are exploring your privacy options, look under "Customize settings" to reset the publics who can see your previously posted photos. Consider restricting your friends' power to "Places you check in to," which means other people can let the world know about your physical location moment by moment. The settings under "Connecting on Facebook" control who can search for you, message you, and make friend requests. "Apps and Websites" is another entire universe of privacy decisions; keep in mind that apps and games often share your personal information with others, including commercial interests. Explore and make conscious decisions about all the options in all the menus as well as submenus that Facebook's Account menu privacy controls give you access to, including but not limited to "information accessible to your friends" and "public search." And more important, if you access Facebook from your laptop or smart phone in public places, "enable https" defends you from people who can steal your passwords and other personal information when you use wi-fi in public.

Understand that you don't have control of what a friend might show a nonfriend. For instance, all my students know that if you want to "Facebook stalk" (my students' term) someone who isn't in your social graph, all you need to do is find out which of your friends is connected to the stalkee, and look over your friend's shoulder at the stalkee's profile. If you publish your genealogy chart, you are providing your and your relatives' mothers' maiden names—frequently a question on password retrieval services—to identity thieves. When you post cute pictures of your children, be aware that you are creating a digital profile for them that will probably never go away.

Facebook is laden with social and political dangers, and Facebook, Inc., isn't in the business of warning people about them. It's up to you to be mindful. And keep in mind that knowing where Facebook's controls are doesn't mean that you are in control. If you want to use Facebook for certain reasons and influence how others can use it to affect you, consider doing what some students have learned to do: deactivate your Facebook account when you log out—every time you log out. You can reactivate it when you log back in, and in the time you are logged off nobody can post messages on your wall, browse your content, or message you. Your social understanding of the norms and boundaries of your friends, networks, and groups increases your social control, but it's crucial to always keep in mind that your control of what Facebook (and other social media) technology can do with as well as to your information and relationships is limited, plus subject to change at any minute.

Facebook was widely used by activists who helped lead the revolutionary changes in the Middle East and North Africa. It has also been used effectively by repressive regimes. One conspiracy theory that isn't quite so crackpot as most is the suspicion that the U.S. Central Intelligence Agency is a secret backer of Facebook because it gives such great access to not only detailed personal data but also the webs of relationships that SNA techniques can reveal. Whether intelligence/law enforcement or commercial interests are behind them, it does appear that someone has created accounts with pictures of attractive women who seem to know everybody, but nobody seems to know them.[73] Or what about this provocative report: "Anonymous, the WikiLeaks-loving online hacktivist collective, claims the US military is developing a piece of software that can infiltrate Facebook and other social networks using an army of fake profiles, cross-referencing information to 'track and identify' individuals."[74] The UK *Guardian*, a thoroughly reputable mainstream publication, confirmed the story.[75]

The world hinted at by these conjectures and exposes is plausible, and worth keeping in mind. The facial recognition feature recently added to Facebook is not a conspiracy theory and certainly could be used for surveillance purposes. In Sudan and Iran, it is definitely not a rumor that the secret police have used Facebook to track and apprehend dissidents.[76, 77] You don't have to believe in the reality of this kind of invasive surveillance state in your particular locality, but it's wise to behave as if it was real. I assume that Web-savvy burglars can cross-reference my status updates and location service check-ins, intelligence agencies know who my friends are, and hundreds of enterprises that want to sell me things have detailed dossiers of my preferences.

I didn't know what to tell my daughter about how to use (and not use) Facebook when she started using it in 2004 (the same year I started teaching). In fact, my daughter introduced me to some of the ins and outs, dos and don'ts, when I joined. The students at Stanford who began using Facebook when I did were the first graduating class to discover that their drunken Facebook photos kept them out of the graduate school or job of their choice. Since then, I've learned from my own experience, discussions with my students (who often choose to use Facebook as the subject of their independent group projects), and researchers, starting with boyd. In particular, boyd notes how "four unique properties of networked publics lead to three important shifts in dynamics":

1. *Persistence*: What you say sticks around. This is great for asynchronous interactions, but not so great when your boss gets to read what you wrote in Usenet back in the 80s.

2. *Replicability*: You can copy/paste from one place to another, taking a conversation from IM and making it available via social network sites. This is also the crux of bullying and how politicians take everything out of context.

3. *Scalability*: The average blog has six readers. Just because things might be public doesn't mean that they automatically will be read by all people across all space and all time. What scales is variable and, often, it's what you least want that is most visible.

4. *Searchability*: My mother would've loved to scream grep into the air and suss out where I had run off to. She couldn't; I'm thankful. Today, through social media, people are tremendously searchable.

These four properties fundamentally shape three different dynamics that alter how people interact in social media environments.

1. *Invisible Audiences*: I can look around here and see who is in the room, but I have no idea who is on the other side of that camera. While I'm at MIT and can make Unix jokes because y'all know what I'm talking about, I have no guarantee that the invisible audience is coming from the same perspective. Online, people have to always negotiate these invisible audiences and that can often be tricky.

2. *Context Collisions*: Part of what makes invisible audiences tricky is that they often represent different social contexts. The properties of networked publics collapse contexts and force people to address social situations with different—and often conflicting—social contexts. This is very tricky.

3. *Public and Private Convergence*: Public and private are usually framed through spatial metaphors with the home being private and everything else being public. Social media confounds this and public and private turn out to be more about control than anything else. Still, it's a matter of collectives and the properties discussed earlier complicate people's ability to control the publicity of any given situation when others around them have a different perspective.

4. *These Dynamics Have Significant Social and Cultural Implications*: They radically alter how people work out identity in relation to those around them. They introduce new structures for social interactions. They complicate power dynamics, agency, and freedom. And yet, finding ways to navigate these spaces means learning how to handle a shifting world with greater complexity. As adults are panicking, teens have been learning. They understand the public world is being radically restructured and they're developing coping mechanisms.[78]

The properties and dynamics that boyd points out are visible in every part of Facebook life if you look at it through boyd's lens. The persistence of all online activities was surprising when Brewster Kahle first started archiving the entire Internet and made the Wayback Machine available— now the irrevocable Web is the norm.[79] When early Internet enthusiasts were plugging modems into telephones, nobody dreamed of a time when every word as well as image ever posted would be stored and searchable forever. If you think you can erase something from the Web nowadays, think again. Invisible audiences have inalterably changed the boundaries between private and public.

The assumption that most of one's social behavior is witnessed by only a few people has to be replaced by the assumption that anything you do online can be witnessed by populations you don't know. Consider the girl in Germany who forgot to check the proper option on the privacy settings of a party announcement and ended up fleeing her own event when fifteen hundred uninvited guests had to be restrained by more than a hundred police officers.[80] Context collisions are what make family holidays so awkward, and why some academics have found themselves in trouble when their naked Burning Man pictures ended up tagged with their name. With video-streaming smart phones all around us, and public surveillance cameras filling in the gaps, we all have to presume that our private actions can become public in a big way at any moment.

Contemplate these four properties and three dynamics when you set up your Facebook profile, update it, and accept or offer friendship, as Facebook forces us to define it. We all experience social pressure and reluctance to hurt people's feelings, but accepting a friend request can affect your behavior and what the world knows about you. Whether you promiscuously accept all friend requests or carefully screen them, know why you are doing what you do. When you give someone access to information about you, you are also giving their network access to it.

Another aspect of Facebook that many Facebook users don't know about is the undisclosed formula that determines which of your friends' updates you see in your News Feed of activities by friends in your social graph. It turns out that you don't always see everybody in your network. Facebook has a secret algorithm that determines what you see. You can influence how the algorithm selects among your friends, but Facebookers, not Facebook, are the ones who will teach you how to do it. To tell Facebook's algorithms that you want to see more of a friend's links, updates, and pictures, you need to visit their profile more often, write on their wall more frequently, and like their updates and links more regularly. Your own links, videos, and photos are more likely to be shared with your friends than your status updates or wall posts, since Facebook is a data-mining enterprise that loves links, videos, and photos. You are also more likely to see people you are tagged with in photos—by you or others.

If you want to have more control, look for the "Edit Options" link on your home page (the News Feed page). This link enables you to set the number of people in your feed; if you increase that number, you'll see more people's feeds. Look for "View Recommended Friends" and click on the names of everyone you want to appear in your News Feed. If you are not comfortable with the facial recognition feature, disable "Suggest photos of me to friends" via "Customize settings" under the account privacy setting method. Look under "Things others share" for an option titled "Suggest photos of me to friends. When photos look like me, suggest my name." Right now, if Facebook has enabled the autosuggestion of photo tags, the "Edit Settings" link next to "Suggest photos of me to friends" will tell you whether or not it is enabled, and allow you to change it to "Disabled" and confirm by clicking "OK." Complicated, isn't it? And don't bet on it getting any simpler over time.

All of this specific advice about settings is subject to change. Facebook has a history of changing its privacy settings and algorithms frequently, and often not informing us of how it has changed our social lives. Like fire, Facebook can be dangerous to you and others in your social network. And like fire, Facebook can be useful. The difference is not only what you know; it's what you keep in mind when you post a link, status update, check-in, or photo, ask or accept a friend connection, like a Web site, or accept a request to play a game or use a new app. At the end of my exploration of today's necessary literacies, the fundamentals of attention literacy are still essential: mindfulness makes a difference, and although the online world is useful and dangerous in part because none of us can control it, each of us has the choice, moment by moment, about whether we are aware of our actions as well as their potential consequences to us personally and our networks.

I've come to the end of my look at attention, crap detection, participation, collaboration, and network literacies. The endnotes and names of cited authorities are included to provide resources for those of you who want to dig deeper and study more broadly. To end, though, I want to consider a thought experiment. For the sake of imagination, let's invert Carr's famous question "Is Google making us stupid?" What value might there be in asking, Used mindfully, how can digital media help us grow smarter? My years of study and experience have led me to conclude that humans are humans because we invent thinking and communicating tools that enable us to do bigger, more powerful things together.

6 How (Using) the Web (Mindfully) Can Make You Smarter

In order to help digital citizens gain skills necessary to succeed online, I've touched on topics that deserve more than the cursory treatment I've been able to devote to them. A few words are essential here, before summarizing the literacies I've presented, about privacy, the public sphere, remix culture, and what parents ought to know about their children's social media use.

Digging into Dataveillance

I've raised privacy concerns in relation to playbor (Do you know who is monetizing your clicks?) and Facebook (Are you sure you know who has access to your profile, pictures, and updates?), but the issue of privacy online far exceeds the implications I've mentioned. Privacy-related issues such as identity theft, state-sponsored surveillance, and behavioral data mining that surmises more about your preferences than you'd prefer anyone to know are the subjects of daily headlines, and touch every aspect of our lives. I've tried to shed some light on the ways our social media usage impacts our private lives. I hope that the knowledge you've gained will give you a greater degree of control over the information others can know about you—and help you know more about the limits of your control over your personal information.

For greater breadth and depth, I recommend George Washington University law professor Jeffrey Rosen's *The Unwanted Gaze: The Destruction of Privacy in America*, and Oxford Internet Institute professor of Internet governance Viktor Mayer-Schönberger's *Delete: The Virtue of Forgetting in the Digital Age*.[1, 2] Mashable, a Web site that has proved to be a reliable source, claims to have an "always up-to-date guide to managing your Facebook privacy."[3]

The time to control dataveillance through policy means has passed. Privacy advocates simply lack the financial and political clout to defeat the privacy-invasion lobbies. While not advocating collective surrender on the legal and judicial front, I do suggest that your best individual defense at the moment is know-how. Find out how to flush cookies that Web sites plant on your computer. Understand the risks when you sign up for a new service, download an app, or accept an invitation to a Facebook game. Figure out where the privacy settings are in the social media you use. You'll still be surveilled. But at least you can be informed.

The Public Sphere

Knowing where the privacy settings can be found isn't enough. You need to understand the political and economic environments in which technology-mediated behavior is embedded. Look closely at the power you have and don't have to control what others know, the people who have power over you because of what they know about you, the ways in which those people and institutions maintain their power, the possibilities for influencing those who have power over you, and the role of media in persuading you and others to believe this or that. This brings us out of the private realm and into the realm of the public.

When you ask whether it is possible for citizens to influence those who have power over us, you are asking about a notion that was formally designated "the public sphere" by Frankfurt school political philosopher Jürgen Habermas.[4] I found out about theories of the public sphere when I wrote *The Virtual Community* in 1991. I asked myself what the most important implications of the then-nascent culture of computer-mediated communication might be. My conclusion was that the issue of political liberty and democratic governance trumped other issues. It doesn't make much sense to quibble over property or privacy if the state can muzzle or murder you for objecting to its policies.

In the eighteenth century, subjects of monarchies in Europe and the Americas did overthrow kings, and then instituted parliaments and constitutions. Habermas's theory of the public sphere was an attempt to explain how this might have happened. If people are to govern themselves and become citizens instead of subjects, they need to have access to accurate information, especially about state policies. Hence the value of a free press (and pressure to control it). It used to be a hanging crime to disclose what was said in Parliament, but at some point transparency in government—initially won through bloody conflict—became institutionalized. Yet the free flow of information to the citizenry is only one side of a bidirectional flow. There must be a way for informed citizens to influence policy and policymakers. This happens, Habermas proposes through the formation of public opinion. Via rational and civil debate about issues, public opinion is formed, and lawmakers pay heed to it. That's the theory at least, in over-simplified form.

The problem, Habermas argued, years before the Internet transformed the nature of broadcast media, was that the well-funded science of public relations was subverting the formation of public opinion by manipulating it. Adorno and Horkheimer, Habermas's predecessors in the Frankfurt school, emphasized the "culture industry" and its "enlightenment as mass deception" in their 1944 book, *Dialectic of Enlightenment*.[5] More recently, others have written about the corruption and consolidation of press ownership along with the deliberate "manufacture of consent."[6, 7] Perhaps the most extreme examples of media-oriented criticism of contemporary democracy are the charges that we all live in a "society of the spectacle" in which popular mass enthusiasms mask the increasingly illusory nature of the world we think we're living in, forming a digitized and hyperrealistic version of Rome's bread and circuses—so realistic that the false world has become a "simulacrum"—an increasingly high-resolution manipulation of perception by the powerful and wealthy.[8, 9]

I cannot easily dismiss these critiques. Anyone who wants to be thoughtful about their use of and by media should be aware of them. But I find the notion that I am the helpless object of psychological manipulation to be disempowering. The critical uncertainty today is whether the radical democratization of access to the means of information production and distribution will change the vectors of influence. Can many-to-many media effectively counter the well-funded disinformation apparatuses of powerful political and economic interests? I raised those questions in *Smart Mobs*. They are still germane; just witness the role of social media in the Arab Spring of 2011. I don't believe that the issue is settled.

Habermas spoke at Stanford a few years ago. I took the occasion to ask him publicly what he thought about his public sphere theory and its subversion by powerful interests, now that the power to broadcast and debate is in the hands of millions. He ducked the question. I blogged about it ("Habermas blows off question about the Internet and the public sphere") and concluded that we need new thinking about this important aspect of our media practices—one that is anchored in but not chained to Habermasian theory.[10] Probably the best short, readable summary of present-day positions about the state of the online public sphere is Pieter Boeder's "Habermas' Heritage: The Future of the Public Sphere in the Network Society" (even though Boeder portrays me as a pom-pom-wielding cheerleader).[11]

I put together an online multimedia minicourse on why the history of the public sphere matters in the Internet age, combining a few short video lectures by myself and others, links to online texts, and a stream of links that I bookmark with the tag "public_sphere."[12] As of this writing, the minicourse has been viewed more than fifty-seven thousand times. At the risk of repeating myself, I believe strongly that what we know and do right now matters. Powerful forces most certainly do influence public opinion. And powerful counterpublics are indeed emerging to challenge power in various forms. To assume that the conflict's outcome is already settled is to surrender to those who seek to control the public sphere, whether they are private interests, corporations, authoritarian states, nihilists, or all of the above. I know my perspective on the online public sphere—specifically, that we can improve it through our actions—has been labeled utopian, and while I certainly agree that the twentieth century taught us to be wary of utopias, I also recall that the rule of law and abolition of slavery were also once utopian ideals.

Pay attention to opportunities you might be given to improve the public sphere. It's not up to anybody else. Apply crap detection when you encounter political assertions, including those you agree with, especially online. Learn to participate in political discussions online and strive to raise the level of debate in the social media public sphere. Contest positions, don't attack people; cite evidence and be willing to change your mind. Collaborate with others to advocate, persuade, and organize; join informed collective action. If you aren't an actor in a democracy, you are the acted on. Know how networks of power and counterpower work, and seek to understand your place in them. The public sphere is a theory about what is, at its base, a simple question: Am I going to act as if citizens acting in concert can wield any power to influence policy? Or am I going to leave my liberty to others?

Remix Ethics and the Politics of Enclosure

An aspect of participation literacy worth consideration, and that ought to be included in at least high-school-level information literacy classes, is "remix culture"—a term derived from Lawrence Lessig's book *Remix: Making Art and Commerce Thrive in the Hybrid Economy* to describe a society that allows and encourages derivative works.[13] Epic conflicts over intellectual property were triggered by the spread of digital technology. Personal computers are printing presses, audio studios, video editors, and copy machines, and the Internet is a worldwide distribution network in which the costs of transporting intellectual property have dropped to nearly nothing. Napster awakened the music recording and motion picture industries to the threat of widespread piracy. When it became trivial to copy a song or movie and upload it, the industries based on buying a physical artifact such as a CD or theater ticket perceived their businesses to be threatened. Millions of people did download and use copyrighted music and video material without paying. Indisputably, this was a form of petty theft that grew to grand larceny when it scaled up. Until Apple's iTunes made it easy for people to purchase music online, the reaction of the recording and motion picture industries to widespread digital theft was to use the legal system to prevent piracy.

Unfortunately, the reaction of incumbent industries went beyond protecting the rights of their intellectual property creators, and by lobbying for the extension of copyright terms as well as laws into the digital realm with the Digital Millennium Copyright Act, incumbent infotainment giants effectively criminalized a vital emerging culture of remixing existing media to convey new meanings, both aesthetic and political. The same legal and political measures also endanger the growth of scientific knowledge through the privatization of scientific publishing, threaten the already-shaky economics of education by making what had previously been considered "fair use" into a licensing maze, and bolster the power of other movements to enclose what had previously been treated as commons (say, the privatization of water, genomes, cell lines, and plants). I raised these issues in my 2002 book *Smart Mobs*, and Lessig covered them incisively in *Free Culture: How Big Media Uses Technology and the Law to Lock Down Culture and Control Creativity*.[14, 15] For a compellingly readable defense of the broader cultural commons and description of the campaign to enclose it, see Lewis Hyde's book from 2010, *Common as Air: Revolution, Art, and Ownership*.[16] And Aram Sinnreich offers a detailed look at the evolution of musical expression in *Mashed Up: Music, Technology, and the Rise of Configurable Culture*.[17]

Remixing isn't just about the disc jockeys who create new forms of art by sampling and remixing music, or video remixers who make statements by juxtaposing news events with popular culture clips to deconstruct power structures and expose political hypocrisy. The arguments Lessig and other "free culture" advocates (and I proudly include myself among them) put forth is that before the rise of multinational intellectual property corporations such as Newscorp and Disney, popular culture, scholarship, and science thrived through a cooperative system in which creators both built on the work of others and made their own work available. It has never been an issue of only public domain or only private profit, but rather of maintaining a balance between the two. In the centuries since copyright was inscribed in the U.S. Constitution, powerful interests have put their thumbs on the scales, causing the balance to shift disastrously toward the private profit end of the spectrum. "Commonists" are not communists. We don't want to abolish private profit from intellectual property; we want a rebalancing that will return a flow of value to everybody, instead of channeling almost all of it to a small number of global culture manufacturers.

The original stated intention of the copyright clause of the U.S. Constitution was to grant to the innovator a temporary monopoly on the reproduction rights to a work ("To promote the Progress of Science and useful Arts, by securing for limited Times to Authors and Inventors the exclusive Right to their respective Writings and Discoverie"), thus repaying the innovator's labors and providing incentive for future innovation. Yet the copyright laws were also intended to enrich the public domain by making those rights available to all after a reasonable period for the innovator to enjoy a monopoly. Over the years, the U.S. Congress—followed by the World Intellectual Property Organization—has extended the copyright term far beyond the lifetime of the original innovators (currently the life of the author plus 70 years, or 120 years after creation for corporate authorship), to the point where established incumbent intellectual property owners ("corporate authors") enjoy a growing advantage, while it becomes more difficult for less powerful writers, musicians, scientists, educators, and filmmakers to make any use of derivative works. As an independent scholar, I encounter these barriers every day when I encounter references—reports of scientific research and scholarly treatises as opposed to high-volume pop-culture works—that are behind paywalls. A researcher seeking to cure cancer can no longer be assured of free access to relevant work by others—not if the commercial sponsors of the research decide to enclose it.

The danger of enclosure is greater than that of stifling cultural expression and slowing the advance of knowledge, important as those threats might be. The conflict over who has the right to use digital media to create and disseminate intellectual property is a war over political control of the power to inform, persuade, educate, debate, and innovate. Arguments about "network neutrality" or licensing of the electromagnetic spectrum might require you to be both a technology geek as well as policy wonk to understand them, but you can be sure that the legislative and judicial decisions that are being made right now will determine whether future innovators will have to ask permission before inventing the World Wide Web or setting up a search engine company in their dorm room. In addition to the books of my own and Lessig's that I have already mentioned, the political dangers of technological enclosure have been highlighted powerfully by Jonathan Zittrain in *The Future of the Internet—and How to Stop It*.[18]

What's a Parent to Do? What's a Parent to Know?

I hope that if you're a parent who has read this far, you now have an expanded view of digital culture. And if you're the parent of a teenager, you understand that in addition to them having fun with their friends and maybe ducking their household chores while they are online, your kids are also creating publics, experimenting with identity, teaching each other technosocial skills, and learning to be active creators of culture. I also presume that any parent diligent enough to read this book will be willing to reconsider the mostly false picture promoted by too much bad journalism that has depicted the Web as a den of frivolity, superficiality, and danger to young people. Teenagers need to experiment with who they are and play with different kinds of identities—and they need to do it with their peers, not just their parents. The public spaces where young people used to hang out have diminished through privatization, surveillance, and prohibition—malls have proliferated while town squares are disappearing; suburbs and urban neighborhoods have few public places where youths are allowed to loiter—so they have created new peer publics in online spaces. What they are learning is not altogether detrimental to themselves and the society they are going to build when they come of age.

For those parents who want to dig deeper, I recommend boyd's dissertation, "Taken Out of Context: American Teen Sociality in Networked Publics," for a sympathetic, informed, and evidence-based examination of exactly what young people are doing online, and why they are doing it.[19] *Hanging Out, Messing Around, and Geeking Out*, quoted in previous chapters, is also eminently readable and evidence based, and offers a positive view of how young people are appropriating social media.[20] I know that *Facebook for Parents*, which Fogg coauthored with Linda Fogg Phillips, is both careful and caring.[21] Sonia Livingstone, professor in the Department of Media and Communications at the London School of Economics and a past president of the prestigious International Communications Association, takes a clear-eyed and empirically based look in *Children and the Internet* at the real as well as imagined risks to youths posed by the Internet.[22] Commonsensemedia.org is an online source of sensible advice and supplies a free curriculum for teaching digital citizenship to youths. Consult these sources and you will be better equipped than almost every other parent in the world to deal with what your teen (or younger) children are doing online.

Because of her expertise in this subject, I asked boyd for advice to parents, and she responded,

When I talk with parents, I encourage them to avoid focusing on the technology, and instead, focus on the underlying issues that worry them as parents. Communication is absolutely essential. As a parent, if you want to help your kids work their way through the challenges brought up because of technology, your best tools are communication, communication, and communication. If something bothers you or you want to know why your child is doing something that you don't understand, ask. In hearing their story, try to understand their perspective first and then offer why you see it differently. For example, you might respond, "But did you think that I might interpret your behavior this way?" Much can be gained by creating a back and forth where you learn how your child is thinking about online activities, and where you share your perceptions. Create a dialogue. The worst thing is when adults come in and say, "Don't do that. This is bad for you. This isn't going to work out." Teens shut down. It's important to create opportunities to share and talk about issues. So a lot of the advice I give to adults is: start listening. When you pay attention to what teens are saying, you'll quickly realize that the challenges most youths face are very similar to challenges that you recognize. Just because they are complicated in new ways by the technologies that are baked into their lives doesn't mean that they are so radically different.[23]

Five Literacies in a Nutshell

Attention

• Social media afford distraction, but attention can be trained.

• Our hormones reward us for information seeking and social contact; they also trigger fight-or-flight physiological responses, although the stimuli that trigger the alarms are not actually threatening. Letting our attention and reactions to social media go uncontrolled can be harmful to our health.

• Breathe! And ask yourself where your attention is directed. These two simple acts, taken together, are your first steps, and powerful levers for bringing your social media attention under control.

• Learning starts with paying attention to others. Let your kids know you are paying attention. Put down the smart phone and look at your children when they address you.

• Most multitasking activities cause you to lose effectiveness in individual tasks. Multitasking, in actuality, usually consists of task switching, and switching comes at a mental cost.

• As Steve Jobs observed, "Focus is about saying no."[24] Making informed decisions about where to deploy your attention begins with realizing that nobody can ever take advantage of all the interesting opportunities the Web presents us. Know that you have to say no, know what you are saying no to, and know why you are doing so.

- Mindfulness and metacognition are about becoming aware of how you are deploying your attention—online and off.
- Breath links mind, brain, and body. Paying attention to your breath helps cultivate mindfulness.
- Attention to intention is how the mind changes the brain. When you repeat mental patterns, you stimulate networks of brain cells and organs to coordinate in ways that strengthen those circuits ("neurons that fire together, wire together"), and make conscious sequences of actions into automatic ensembles—like learning to read—except that now you are learning to manage attention in the face of social media opportunities for distraction. To say that social media affords distraction is not to say that the technology has to be in control of your awareness; you can take control back, through repeated, conscious efforts.
- Breathe! (It's worth repeating.) And while you are examining how you are deploying your attention, take the occasion to check in with your body. Don't remain seated for hours. Your attention and body shouldn't drift apart for too long without reanimating yourself. In addition to breathing and knowing it, stand up and move regularly.
- Social media attention training requires understanding your goals and priorities (intentions), and involves asking yourself, at regular intervals, whether your current activity at any moment moves you closer to your goal or serves your higher priorities (attention).
- Like meditation, mindful use of social media begins with noticing when your attention has wandered, and then gently bringing it back to focus on your highest priority—like training a puppy.
- Sometimes your immediate goals are to learn or make friends, in which case, having fun, making small talk, and exploring from link to link is not just permissible, it's important. Don't make attention training a war on flow.
- To establish new attention habits, start small, find a place in your routine for a new behavior, and repeat until paying attention has become habitual.

Crap Detection

- The answer to almost any question is available, if you know how to search.
- When searching, think about what words might be on the page you seek; think about possible answers when posing the question. Add terms like "how to" or "critique" to find instruction or alternative views.

• Don't stop with one search if you are learning about a topic and not just looking for the nearest pizza parlor. Regard search as a process of investigation. Sometimes, instead of searching to find, search to discover.

• Look at the third, fourth, and fifth pages of search results. Formulate new searches based on the terms you find in relevant snippets from previous searches.

• Understand that it is up to you to determine whether the result of your search or material you find online in any other way is accurate, inaccurate, or intentionally misleading.

• Start out skeptical, and then by "thinking like a detective," verify information for yourself.

• Look for an author. And search on the author's name. Use Whois and other tools to look behind the surface of a Web site.

• Look for sources if a site makes assertions. Find out what others say about those sources.

• Use the search term "link: http:// . . . " (with your URL in place of the ellipses) to see every link to a specified page.

• To check claims by any political faction, use factcheck.org. Know how to check urban legend and medical hoax sites. The crap-detection equivalent of infotention is to create your own dashboard of credibility-checking tools, and teach yourself to use the appropriate tool quickly.

• When conducting medical searches, make use of resources mentioned in this book—such as the Health on the Net Foundation—and the resources that they recommend. Trust can be transitive if, and only if, your root source is maximally trustworthy.

• When you see breaking news via social media, triangulate: try to find three sources to verify before passing along a rumor.

• Be aware of filter bubbles and echo chambers. Question opinions you agree with and pay attention to sources you disagree with. The search engine DuckDuckGo claims to be able to help you break out of your search filter bubble.[25]

• Infotention combines attentional discipline with information-handling skills. Use dashboards, filters, and radars, and arrange the spatial organization of your dashboards to reflect your priorities.

• Learn to make rapid microdecisions about whether to pay attention to information at all, open a browser tab for later, and bookmark and/or curate it. Start out deliberately, weighing the potential distraction against the goals you've decided on, and work to make the decision-making process more automatic.

Participation

• Every PC as well as smart phone is a printing press, broadcasting station, political organizing tool, and site for growing a community or marketplace. Knowledge, power, advantage, companionship, and influence lie with those who know how to participate, rather than those who just passively consume culture.

• The Web's architecture of participation enables you (and everybody else) to act in your own self-interest in ways that create value for everybody. Participation points both inward and outward: it's about personal empowerment, and also building things together.

• We're in the early years of the emergence of a participatory culture. People who think of themselves as capable of creating as well as consuming are different kinds of citizens, and our collective actions add up to a different kind of society.

• Participation can start with lightweight activities such as tagging, liking, bookmarking, and wiki editing, then move to higher engagement with curation, commenting, blogging, and community organizing.

• Curation is a form of participation that can refine your captured information into contextualized knowledge, enhance your reputation, and serve the information needs of others at the same time.

• Awareness of your digital footprints and impacts of your digital profiles ought to precede your conscious participation online. Think before you post, because your digital actions are findable, reproducible, and available to people you don't know, and will remain available to all indefinitely.

• Learning the norms and boundaries of an online culture is the first step in participation, whether in a virtual community or online subculture.

• Assume goodwill; when you think someone else is attacking you, it's likely a misunderstanding. Ask politely for clarification.

• Irony and sarcasm don't go over well in text-only communications. The harshness of possible interpretations is ordinarily modulated by tone of voice, facial expression, and body language.

• Sometimes when you are having fun online, someone else is profiting. It helps to be aware of who might be exploiting your playbor.

• Crap detect thyself before broadcasting something as an assertion of fact.

Collaboration

• Humans are human because we use communication to organize collective action. Social media can amplify collective action.

• Although the dominant narrative of biology, economics, and daily life has stressed competition and conflict, recent findings indicate a much larger role for cooperative arrangements.

• You need coordination to dance by yourself, collaboration to dance with a partner, and cooperation to organize a dancing flash mob.

• Coordination is the simplest—like adjacent rice farmers flooding their fields at the same time to control pests.

• Cooperation involves coordination, but it also includes sharing resources and acting in concert toward shared interests.

• Collaboration requires agreement about shared goals. Everyone can look after their own interests, but communication and negotiation are required for sharing goals.

• When crowdsourcing or soliciting volunteers in mass collaboration, be humble, offer a wide variety of ways to contribute, encourage self-election, recognize contributors, and provide contributors with ways to communicate with each other.

• To host a virtual community, make the rules clear at the beginning, strive for the participants to gain a sense of ownership, and act the way a host at a good party does—welcome people, introduce them, break up fights, and model the kind of behavior you would like to see.

• Digital media and networks are new means of economic production and distribution that make possible social production as well as the firm and market. Volunteers can now create significant value—software, encyclopedias, citizen science, open educational resources, and products that haven't been invented yet.

• People contribute to nonmarket social production in order to learn, gain reputation, meet others, and contribute to a common good.

• People are using social media to consume as well as create collaboratively, setting off a "rapid explosion in traditional sharing, bartering, lending, trading, renting, gifting, and swapping reinvented through network technologies on a scale and in ways never before possible."[26]

• A group's collective intelligence is correlated with neither the average or highest IQ of its members but instead with the members' diversity and facility at conversational turn taking.

• Casual conversation builds trust and enables more instrumental collaboration in all forms of networked collaboration.

Network Smarts

• Networks have structures that influence the way individuals and groups behave.

• Small-world networks allow information to travel rapidly across large populations. A small number of random, distant connections are required. Small-world networks permeate the world. Human sociality, chemical reactions in living cells, and electric transmission grids share key properties.

• In small-world networks, a small number of hubs will have a large number of connections. Most nodes will be out in the long tail with smaller numbers of connections. In online social networks, content and value can climb the long tail to the heavily trafficked hubs.

• Networks that enable many-to-many communication grow in value more rapidly than broadcast networks. Networks that enable group formation grow in value more rapidly than nongroup-forming many-to-many networks.

• Human social networks maintained through the medium of speech go back to the origin of our species. Technologically networked communication media extend and amplify the reach of traditional social networks to make new forms of sociality possible.

• Online networks that support social networks share properties of more general network structure as well as the specific properties of human networks.

• Social media are permitting people to seek support, information, and a sense of belonging from sparsely knit, loosely bound networks as well as the traditional densely knit, tightly bound groups.

• A portfolio of both strong and weak ties is useful to individuals in a network society. Social media can help maintain larger networks of weak ties, but only a limited number (around 150) of strong-tie relationships can be maintained, irrespective of media. Social media make it possible to maintain ongoing ties with people whose social networks you have left through graduation, moving, or changing jobs.

• One's position in social networks matters. Centrality—how many people and networks go through you to connect with each other—can be more important than the number of connections.

• Diverse networks are collectively smarter and provide a richer variety of resources to participating nodes.

• People who can bridge networks—fill structural holes—stand to benefit.

• Social capital grows for individuals and groups from networks of trust and norms of reciprocity. Bonding capital enables trust between individuals. Bridging capital connects individuals and groups into larger networks.

• Small talk nourishes trust. Trust lubricates transaction.

• Pay it forward. Doing favors for others online is the strongest predictor of whether you will receive favors from others.

• Facebook privacy settings require attention. Make sure you know what you are sharing, and with whom you are sharing it.

• Facebook and other social network services have four unique properties that lead to three crucial shifts in dynamics.[27] They are persistent, replicable, scalable, and searchable, and thus lead to colliding contexts, invisible audiences, and public and private convergences.

Literacy as Lever, as Divide

As I noted at the outset of our learning journey, the emerging digital divide is between those who know how to use social media for individual advantage and collective action, and those who do not. Throughout this exploration of the five literacies I consider to be essential, I've stressed both personal empowerment and contributions to the commons. That's how social media *can* work: we can better ourselves and also improve the environment in which everybody's attention dwells at least part of each day. But the effectiveness of institutions for collective action—the social levers that humans have created to balance self-interest and public goods—depends on people acting in concert.

Just as social media may have been necessary, but not sufficient enabling conditions for the revolutionary changes in North Africa and the Middle East known as the Arab Spring, a social media commons in which information is useful and trustworthy, discussions are civil and productive, and networked collaborations generate social capital is possible yet not guaranteed. It took brave and dedicated people to overthrow autocratic regimes through people power. It will take dedicated and knowledgeable people to spread the word about social media know-how.

Educational institutions cannot change swiftly and broadly enough to match the pace of change in digital culture—and public education faces other problems. I do hope that at least some simple form of crap detection finds its way into K–12 classrooms. At the same time that institutions are struggling to react, networks of gamers, Wikipedians, Tumbloggers, and other virtual communities of peer-to-peer learners are experimenting with new ways to spread literacies—expressive and interpretative skills with a social element, grounded in a shared context. Research indicates that the factor most likely to positively influence a young person's ability to distinguish good from bad information online is their parents' education level; significant socioeconomic divides have not gone away online.

But the good news is that knowledge and know-how can spread through online networks as swiftly as well as pervasively as a viral video. And it's up to us to do so. Let's work small-world networks to our advantage. Spread some of what you've learned through your own networks and see if you can inject it into disparate networks you might connect to. Telling people about this book is one way to do that. So is practicing what you've learned.

I chose the five literacies I thought were the most important right now, but that doesn't mean that these are the only bodies of knowledge and social contexts that are critical and worth learning. Scientific literacy, for example—understanding the basics of how scientific inquiry works—is increasingly needed, especially now that strong political forces seem determined to roll back the Enlightenment values of reason and empiricism. Knowing how to program, as Douglas Rushkoff notes in *Program or Programmed: Ten Commands for a Digital Age*, is too important to be left to any elite.[28] Information and media literacies both overlap with the five literacies I've introduced, and a search-to-discover on either term will reveal rich networks of people and resources. Social and emotional intelligence can also be seen as literacies. The proliferation of literacies and divides that accompany them are a real problem. It isn't easy to maintain a high level of basic reading and writing literacy, and the percentage of the population that can afford the time and money to learn additional multiple literacies is undoubtedly going to remain small, but that doesn't mean it has to be an elite. The multiliterate can be a public—a networked public.

We are only beginning to see what networked publics can do for good and evil. I have chosen to try to provide resources to increase the amount of good that networked publics can do. I don't claim that this is a sufficient solution to the problem of proliferating literacies and publics. I have been accused of being an optimist, which I am not. I am aware that the deck is always stacked by those who have the most at stake if they can manage a way to do it. Nevertheless, I choose to be hopeful. We are all descended from predecessors who, while their companions might become realistically resigned to the hopelessness of their situation, couldn't help thinking, "There must be some way out of this." The future is not guaranteed. There is no influence without knowledge and effort. I've tried to provide tools for you to gain that knowledge. It's up to you to make the effort.

Notes

Introduction

1. For a look at the influence of the early use of technologies, see Langdon Winner, "Do Artifacts Have Politics," in *The Whale and the Reactor: A Search for Limits in an Age of High Technology*, ed. Langdon Winner (Chicago: University of Chicago Press, 1986), 19–39. I have adopted Winner's use of the word regime to describe the constellation of physical mechanisms, supply and waste infrastructure, political regulation, economic exploitation, cultural norms, social benefits, and negative impacts that accompany the widespread adoption of a technology. I learned this use of the word from Winner when I asked him to contribute to a *Whole Earth Review* edition devoted to "Questioning Technology" (I was editor in chief at that time); see Langdon Winner, "Artifacts/Ideas and Political Culture," *Whole Earth Review* 73 (winter 1991): 18–26.

2. Tim O'Reilly, "The Architecture of Participation," 2004, http://oreilly.com/pub/a/oreilly/tim/articles/architecture_of_participation.html.

3. Elizabeth Eisenstein, *The Printing Press as an Agent of Change: Communications and Cultural Transformations in Early Modern Europe*, 2 vols. (Cambridge: Cambridge University Press, 1979).

4. Claude S. Fischer, *America Calling: A Social History of the Telephone to 1940* (Berkeley: University of California Press, 1994).

5. "Gutenberg's Legacy," Harry Ransom Center, University of Texas at Austin, http://www.hrc.utexas.edu/educator/modules/gutenberg/books/legacy.

6. The phrase is from Mizuko Ito, *Hanging out, Messing Around, and Geeking Out: Kids Living and Learning with New Media* (Cambridge, MA: MIT Press, 2009).

7. Howard Rheingold, *Tools for Thought* (New York: Simon and Schuster, 1985).

8. Howard Rheingold, "Virtual Communities," *Whole Earth Review* 57 (winter 1987): 78–82 reprinted in *Journal of Virtual Worlds Research* 1, no. 1 (July 2008), http://journals.tdl.org/jvwr/article/view/293/247.

9. Cathy Davidson, "This Is Your Brain on the Internet, Part 2," blog submission, January 9, 2009, http://www.hastac.org/node/1926.

10. See http://www.henryjenkins.org/2006/10/confronting_the_challenges_of.html for Jenkins and http://networkedpublics.org/book/introduction for Ito.

11. Rheingold, "Virtual Communities," 3.

12. Available at http://www.patientslikeme.com and http://www.mdjunction.com.

13. Yochai Benkler, *The Wealth of Networks: How Social Production Transforms Markets and Freedom* (New Haven, CT: Yale University Press, 2006), 63.

14. "The Library of Congress' Photostream," last modified February 11, 2011, http://www.flickr.com/people/library_of_congress.

15. Steve Silberman, "Inside the High Tech Search for a Missing Silicon Valley Legend," *WIRED* 15, no. 8 (July 2007), 130–139, 154–155.

16. Manuel Castells, "Why Networks Matter," in *Network Logic: Who Governs in an Interconnected World?* ed. Helen McCarthy, Paul Miller, and Paul Skidmore, 221–224 (London: Demos, 2004).

17. Nicholas Carr, "Is Google Making Us Stupid?" *Atlantic*, July–August 2008, 56–63; and Carr, *The Shallows* (New York: W. W. Norton, 2010), 64.

18. Mary Madden and Lee Rainie, "Adults and Cell Phone Distractions," report data set, May 2010 cell phones (Washington, DC: Pew Internet and American Life Project, June 2010), http://pewinternet.org/Reports/2010/Cell-Phone-Distractions.aspx.

19. Amanda Lenhart, Mary Madden, and Paul Hitlin, "Teens and Technology: Youth Are Leading the Transition to a Fully Wired and Mobile Nation" (Washington, DC: Pew Internet and American Life Project, July 2005), http://www.pewinternet.org/~/media/Files/Reports/2005/PIP_Teens_Tech_July2005web.pdf.

20. Amanda Lenhart and Mary Madden, "Teen Content Creators and Consumers" (Washington, DC: Pew Internet and American Life Project, November 2005), http://www.pewinternet.org/Reports/2005/Teen-Content-Creators-and-Consumers.aspx.

21. Jay Rosen, "The Legend of Trent Lott and the Weblogs," *Pressthink Weblog*, March 2004, http://archive.pressthink.org/2004/03/15/lott_case.html; Howard Kurtz, "After Blogs Got Hits, CBS Got a Black Eye," *Washington Post*, September 20, 2004, http://www.washingtonpost.com/wp-dyn/articles/A34153-2004Sep19.html; "George Allen Introduces Macaca," uploaded by zkman, August 15, 2006, http://www.youtube.com/watch?v=r90z0PMnKwI.

22. "The End of Universal Rationality: A Talk with Yochai Benkler," interview, *Edge*, March 31, 2009, http://www.edge.org/documents/archive/edge279.html.

23. South-East Asia Earthquake and Tsunami Blog, last modified August 11, 2009, http://tsunamihelp.blogspot.com.

24. R.I.M. Dunbar, "Coevolution of Neocortical Size, Group Size, and Language in Humans," *Behavioral and Brain Sciences* 16, no. 4 (1993): 681–735.

25. Howard Rheingold, "A Slice of Life in My Virtual Community," June 1992, http://www2.fiu.edu/~mizrachs/virt-comm.html.

26. Available at, respectively, http://www.diigo.com; http://www.delicious.com; http://www.quora.com; http://www.formspring.me.

27. Eric Steven Raymond, "The Origins and History of Unix, 1969–1995," in *The Art of Unix Programming* (Boston: Addison-Wesley, 2003), http://www.faqs.org/docs/artu/ch02s01.html.

28. Howard Rheingold, "The Loneliness of the Long-distance Thinker," in *Tools for Thought: The History and Future of Mind-Expanding Technology* (New York: Simon and Schuster, 1985), 174.

29. Stanley Milgram, "The Small World Problem," *Psychology Today* 2 (1967): 60–67.

30. Duncan J. Watts, *Six Degrees: The Science of a Connected Age* (New York: W. W. Norton, 2003).

31. Nicholas Christakis and James Fowler, *Connected: The Surprising Power of Social Networks* (Boston: Little, Brown, 2009).

32. David P. Reed, "That Sneaky Exponential: Beyond Metcalfe's Law to the Power of Community Building," *Context* (Spring 1999), http://www.reed.com/dpr/locus/gfn/reedslaw.html.

33. Castells, "Why Networks Matter."

34. "About Us," fliers' rights, last modified March 31, 2011, http://flyersrights.org/about.php; Eli Lake, "Hacking the Regime: How the Falung Gong Empowered the Iranian Uprising," *New Republic*, September 3, 2009, 18–20.

35. Robin Dunbar, *Grooming, Gossip, and the Evolution of Language* (Cambridge, MA: Harvard University Press, 1997), 79.

36. Conversation with danah boyd, June 15, 2010, via Skype.

37. Robert Leonardi, Rafaella Y. Nanetti, and Robert Putnam, *Making Democracy Work: Civic Traditions in Modern Italy* (Princeton, NJ: Princeton University Press, 1993), 163.

38. "Charity: Water," Twestival, last modified February 12, 2009, http://www.charitywater.org/twestival.

39. Douglas Engelbart, "Augmenting Human Intellect: A Conceptual Framework," Stanford Research Institute, October 1962, http://www.invisiblerevolution.net/engelbart/full_62_paper_augm_hum_int.html; J. C. R. Licklider, "Topics for Discussion at the Forthcoming Meeting, Memorandum For: Members and Affiliates of the Intergalactic Computer Network," Washington, DC, Advanced Research Projects Agency, April 23, 1963, http://www.kurzweilai.net/articles/art0366.html.

40. Rheingold, *Tools for Thought*, 150. (For an account that sheds light on the military-industrial institutions and forces that also contributed to the evolution of PCs, see Paul Edwards, *The Closed World: Politics and Discourse in Cold War America* (Cambridge, MA: MIT Press, 1996).

41. Peter F. Drucker, "Knowledge-Worker Productivity: The Biggest Challenge," *California Management Review* 41, no. 2 (Winter 1999), 79–94, http://rfrost.people.si.umich.edu/courses/527-1/Drucker2.pdf.

42. Ted Nelson, *Computer Lib: You Can and Must Understand Computers Now / Dream Machines: New Freedoms through Computer Screens—A Minority Report*, 1st ed. (South Bend, IN: the distributors, 1974), 159; Fred Turner, *From Counterculture to Cyberculture: Stewart Brand, the Whole Earth Network, and the Rise of Digital Utopianism* (Chicago: University of Chicago Press, 2006), 133.

43. Dineh Moghadam, "The New York Times Information Bank in an Academic Environment and a Computer-Assisted Tutorial for its Non-Specialist Users" (PhD diss., University of Pittsburgh, 1974), http://www.eric.ed.gov/ERICWebPortal/detail?accno=ED135373.

44. Alan Kay, "Microelectronics and the Personal Computer," *Scientific American*, September 1977, 230–244, http://www.guidebookgallery.org/articles/microelectronicsandthepersonalcomputer.

45. Ibid., 230.

46. Michael A. Hiltzik, *Dealers of Lightning: Xerox PARC and the Dawn of the Computer Age* (New York: Harper Business, 2001).

47. Engelbart, "Augmenting Human Intellect."

48. "The Demo," MouseSite, http://sloan.stanford.edu/mousesite/1968Demo.html.

49. Engelbart, "Augmenting Human Intellect."

50. Rheingold, "Virtual Communities."

Chapter 1

1. "Attention in the Classroom," video clip, uploaded by howardrheingold, August 27, 2010, http://blip.tv/howardrheingold/attention-in-the-classroom-4073002.

2. Clifford Nass, Eyal Ophir, and Anthony D. Wagner, "Cognitive Control in Media Multitaskers," in *Proceedings of the National Academy of Sciences*, vol. 106, no. 37 (April 2, 2009), 15583–15587.

3. Matt Richtel, "Only a Few Can Multitask," *New York Times Bits Blog*, March 30, 2010, http://bits.blogs.nytimes.com/2010/03/30/only-a-few-can-multi-task.

4. Jason M. Watson and David L. Strayer, "Supertaskers: Profiles in Extraordinary Multitasking Ability," *Psychonomic Bulletin and Review* 7, no. 4 (2010): 479–485.

5. "Study Finds Fighter Pilot Brains Are Unique," *RedOrbit*, December 15, 2010, http://www.redorbit.com/news/health/1966900/study_finds_fighter_pilot_brains_are_unique/index.html.

6. Heleen A. Slagter, Antoine Lutz, Lawrence L. Greischar, Andrew D. Francis, Sander Nieuwenhuis, James M. Davis, Richard J. Davidson, "Mental Training Affects Distribution of Limited Brain Resources," *PLoS Biology* 5, no. 6, http://www.plosbiology.org/article/info:doi/10.1371/journal.pbio.0050138.

7. George A. Miller, "The Magical Number Seven, Plus or Minus Two," *Psychological Review* 63 (1956): 81–97.

8. P. Sven Arvidson, "A Lexicon of Attention: From Cognitive Science to Phenomenology," *Phenomenology and the Cognitive Sciences* 2, no. 2 (2003): 99–132.

9. Maggie Jackson, "May We Have Your Attention, Please?" *Bloomberg Businessweek*, June 12, 2008, http://www.businessweek.com/magazine/content/08_25/b4089055162244.htm.

10. David G. Myers, *Psychology* 7th ed. (Holland, MI: Worth Publishers, 2004).

11. Giacomo Rizzolatti and Laila Craighero, "The Mirror-Neuron System," *Annual Review of Neuroscience* 27 (2004): 169–192.

12. Stanislas Dehaene, *Reading in the Brain: The Science and Evolution of a Human Invention* (New York: Viking, 2009), 149.

13. I recommend reading it. Cathy Davidson, *Now You See It: How the Brain Science of Attention Will Change the Ways We Live, Work, and Learn* (New York: Viking 2010).

14. Cathy Davidson, "Here's How the Brain Science of Attention Really Works," *HASTAC Weblog*, October 9, 2010, http://www.hastac.org/blogs/cathy-davidson/heres-how-brain-science-attention-really-works.

15. Ibid.

16. Nelson Cowan, Emily M. Elliot, J. Scott Saults, Candace C. Morey, Sam Mattox, and Anna Hismjatullina, and Andrew R. A. Conway. "On the Capacity of Attention: Its Estimation and Its Role in Working Memory and Cognitive Aptitudes," *Cognitive Psychology* 51 (2005): 42–100.

17. Barry Arons, "A Review of the Cocktail Party Effect," MIT Media Lab, http://xenia.media.mit.edu/~barons/html/cocktail.html.

18. "Selective Action Test," uploaded by profsimons, March 10, 2010, http://www.youtube.com/watch?v=vJG698U2Mvo.

19. Maggie Jackson, *Distracted: The Erosion of Attention and the Coming Dark Age* (Amherst, NY: Prometheus Books, 2008).

20. Nicholas Carr, "Is Google Making Us Stupid?" *Atlantic*, July–August 2008: 56–63.

21. Alison Gopnik, "Diagnosing the Digital Revolution," *Slate*, February 7, 2011, http://www.slate.com/id/2283467.

22. Linda Stone, "Just Breathe: Building the Case for Email Apnea," *Huffington Post*, February 8, 2008, http://www.huffingtonpost.com/linda-stone/just-breathe-building-the_b_85651.html.

23. Ibid.

24. Matt Richtel, "Driven to Distraction: Drivers and Legislators Dismiss Cellphone Risks," *New York Times*, July 19, 2009, A1.

25. Matt Richtel, "In Study, Texting Lifts Crash Risk by Large Margin," *New York Times*, July 27, 2009, A1.

26. David L. Strayer, Frank A. Drews, and Dennis L. Crouch, "A Comparison of the Cell Phone Driver and the Drunk Driver," *Human Factors: The Journal of the Human Factors and Ergonomics Society* 48, no. 2(Summer 2006): 381–391.

27. Tony Schwartz, "Breaking the Email Addiction," *Harvard Business Review*, June 29, 2010, http://blogs.hbr.org/schwartz/2010/06/breaking-the-email-addiction.html.

28. Atsunori Ariga and Alejandro Lleras, "Brief and Rare Mental 'Breaks' Keep You Focused: Deactivation and Reactivation of Task Goals Preempt Vigilance Decrements," *Cognition* 118, no. 3 (2011): 439–443.

29. Matt Richtel, "Attached to Technology and Paying a Price," *New York Times*, June 6, 2010.

30. Emily Salvaterra, K. T. Hanson, Gonzalo Brenner, and Hannah Jackson, "Caught in the Net: The Internet and Compulsion," *Neuroanthropology Blog*, May 28, 2009, http://neuroanthroplogy.net/2009/05/28/caught-in-the-net-the-internet-compulsion.

31. Youmasu J. Siewe, "Internet Addiction: When the Computer Becomes a Problem," *Stillwater News Press*, June 9, 2009, http://forum.psychlinks.ca/internet-behavior/17928-internet-addiction-when-the-computer-becomes-a-problem.html.

32. Richtel, "Attached to Technology, A1."

33. Adam L. Penenberg, "Social Networking Affects Brain Like Falling in Love," *Fast Company*, July 1, 2010, 78–113.

34. Ibid.

35. Ellen Reagan, "Stuck in the 'BlackBerry Zone': Kids Feel Neglected by Media-Obsessed Parents, But Some Trying to Change Ways," *Chicago Sun-Times*, November 11, 2010, S10.

36. Cited in ibid.

37. Daniel J. Siegel, *The Mindful Brain: Reflection and Attunement in the Cultivation of Well-being* (New York: W. W. Norton, 2007), 26.

38. Cited in Katie Hafner, "Texting May Be Taking a Toll," *New York Times*, May 26, 2009, D1.

39. Cited in Reagan, "Stuck in the 'BlackBerry Zone.'"

40. Sherry Turkle, interview with the author, January 3, 2011.

41. Cited in Sharon Jayson, "2010: The Year Technology Replaced Talking," *USA Today*, June 30, 2010, http://www.usatoday.com/yourlife/parenting-family/2010-12-30-1AYEAR30_CV_N.htm. See also Claude Fischer, *America Calling: A Social History of the Telephone to 1940* (Berkeley: University of California Press, 1991).

42. Sherry Turkle, *Alone Together: Why We Expect More from Technology and Less from Each Other* (New York: Basic Books, 2011).

43. Cited in Jayson, "2010."

44. Turkle, interview.

45. Ibid.

46. Carr, "Is Google Making Us Stupid?": 60.

47. Ibid., 60.

48. Carr, *The Shallows*, 118, 138.

49. Ibid.

50. Cited in Gregory M. Lamb, "Are iPads, Smartphones, and the Mobile Web Rewiring the Way We Think?" *Christian Science Monitor*, July 24, 2010, http://www.csmonitor.com/Innovation/Tech/2010/0724/Are-iPads-smartphones-and-the-Mobile-Web-rewiring the way-we-think.

51. Clay Shirky, *Encyclopaedia Britannica Blog*, July 17, 2008, http://www.britannica.com/blogs/2008/07/why-abundance-is-good-a-reply-to-nick-carr.

52. Naomi S. Baron, *Always On: Language in an Online and Mobile World* (New York: Oxford University Press, 2008), 5.

53. Andrew Keen, *The Cult of the Amateur: How Today's Internet Is Killing Our Culture* (New York: Crown, 2007).

54. Cited in Meredith Melnick, "Is Technology Making Us Lonelier?" *Time*, January 10, 2011, http://www.time.com/health/article/0,8599,2041714,00.html.

55. Maggie Jackson, *Distracted: The Erosion of Attention and the Coming Dark Age* (Amherst, NY: Prometheus Books, 2008), 13.

56. Ibid, 14.

57. Ibid, 34.

58. Maggie Jackson, "Attention Class," Boston.com, June 29, 2008, http://www .boston.com/news/education/higher/articles/2008/06/29/attention_class.

59. Steve Lohr and Miguel Helft, "Google Gets Ready to Rumble with Microsoft," *New York Times*, December 16, 2007, 31.

60. Linda Stone, "Beyond Simple Multi-tasking: Continuous Partial Attention" (November 30, 2009), http://lindastone.net/2009/11/30/beyond-simple-multi-task-ing-continuous-partial -attention.

61. Linda Stone, interview with the author, August 9, 2010.

62. Maryanne Wolf, *Proust and the Squid: The Story and Science of the Reading Brain* (New York: HarperCollins, 2007), 3.

63. Ibid., 5.

64. Dehaene, *Reading*.

65. Wolf, *Proust and the Squid,* 12.

66. Steven Pinker, foreword to *Why Our Children Can't Read—and What We Can Do about It: A Scientific Revolution in Reading*, by Diane McGuinness (New York: Simon and Schuster, 1997), ix; cited in Wolf, *Proust and the Squid*, 19.

67. Dehaene, *Reading*, 121.

68. Wolf, *Proust and the Squid*, 71.

69. Mark Whipple, "The Dewey-Lippmann Debate Today: Communication Distortions, Reflective Agency, and Participatory Democracy," *Sociological Theory* 23, no. 2 (June 2005), 156–178.

70. Wolf, *Proust and the Squid*, 221.

71. Ibid., 226.

72. Soren Gordhamer, "The LAKERS Meditate?" Mind-Body Awareness Project, May 17, 2006, http://www.mbaproject.org/4Youth/the-chicago-bulls-meditate-so-do-the -lakers.

73. Gary Wolf, "What It Is, Is Up to Us," *Reed Magazine*, February 2002, 10–13.

74. Joe Kamiya, "Operant Control of the EEG Alpha Rhythm and Some of Its Reported Effects on Consciousness," in *Biofeedback and Self-Control: An Aldine Reader on the Regulation of Bodily Processes and Consciousness* (Piscataway, NJ: AldineTransaction, 1971).

75. Lester Fehmi and Jim Robbins, *Open Focus: Harnessing the Power of Attention to Heal Mind and Body* (New York: Trumpeter, 2007).

76. Steven Johnson, *Mind Wide Open: Your Brain and the Neuroscience of Everyday Life* (New York: Scribner, 2004).

77. Jon Kabat-Zinn, "Mindfulness-based Interventions in Context: Past, Present, and Future," *Clinical Psychology: Science and Practice* 10, no. 2 (June 2003): 145.

78. Soren Gordhamer, *Wisdom 2.0: Ancient Secrets for the Creative and Constantly Connected* (New York: HarperCollins, 2008). I highly recommend this book.

79. Gordhamer, "The LAKERS Meditate?" http://www.mbaproject.org/4Youth/the-chicago-bulls-meditate-so-do-the-lakers.

80. Jennifer A. Livingston, "Metacognition: An Overview," http://gse.buffalo.edu/fas/shuell/cep564/metacog.htm.

81. "Metacognition," Wikipedia, http://en.wikipedia.org/wiki/Metacognition.

82. Livingston, "Metacognition."

83. Norman A. Farb, Zindel V. Segal, Helen Mayberg, Jim Bean, Deborah McKeon, Zainab Fatima, and Adam K. Anderson, "Attending to the Present: Mindfulness Meditation Reveals Distinct Neural Modes of Self-reference," *Social Cognitive and Affective Neuroscience* 2, no. 4 (June 23, 2007): 313–322.

84. David Rock, "The Neuroscience of Mindfulness," *Psychology Today*, October 11, 2009, http://www.psychologytoday.com/blog/your-brain-work/200910/the-neuroscience-mindfulness.

85. Amishi P. Jha, Jason Krompinger, and Michael J. Baime, "Mindfulness Training Modifies Subsystems of Attention, *Cognitive and Affective Behavioral Neuroscience* 7, no. 2 (June 2007), 109.

86. Thom Shanker and Matt Richtel, "In New Military, Data Overload Can Be Deadly," *New York Times*, January 16, 2011, A1.

87. Michael I. Posner and Brenda Patoine, "How Arts Training Improves Attention and Cognition," Dana Foundation, September 14, 2009, http://www.dana.org/news/cerebrum/detail.aspx?id=23206.

88. Pamela D. Hall, "The Effect of Meditation on the Academic Performance of African American College Students," *Journal of Black Studies* 29, no. 3 (January 1999): 408–415.

89. Cited in Jonah Lehrer, "Don't! The Secret of Self-control," *New Yorker*, May 18, 2009, 27.

90. Ibid., 29.

91. Cited in ibid., 27.

92. Norman Doidge, *The Brain That Changes Itself: Stories of Personal Triumph from the Frontiers of Brain Science* (New York: Penguin, 2007), 63.

93. Ole Paulsen and Terrence J. Sejnowski, "Natural Patterns of Activity and Long-term Synaptic Plasticity," *Current Opinion in Neurobiology* 10, no. 2 (2000): 172–179, http://www.ncbi.nlm.nih.gov/pubmed/10753798.

94. Siegel, *The Mindful Brain*, 176.

95. Stephen LaBerge and Howard Rheingold, *Exploring the World of Lucid Dreaming* (New York: Ballantine, 1988).

96. Richard Seven, "Life, Interrupted," *Pacific Northwest: The Seattle Times Magazine*, November 28, 2003, http://seattletimes.nwsource.com/pacificnw/2004/1128/cover.html.

97. David Levy, "Syllabus: Information and Contemplation," Information School, University of Washington, spring quarter 2006, http://www.contemplativemind.org/programs/academic/syllabi/levy.pdf.

98. David Levy, "No Time to Think," Google tech talk, March 5, 2008, http://www.youtube.com/watch?v=KHGcvj3JiGA.

99. Linda Stone, "Continuous Partial Attention," http://lindastone.net/qa/continuous-partial-attention.

100. Linda Stone, "Is It Time to Retire the Never-ending List?" *Huffington Post*, June 11, 2008, http://www.huffingtonpost.com/linda-stone/is-it-time-to-retire-the_b_106624.html.

101. Peter Bregman, "An 18-Minute Plan for Managing Your Day," *Harvard Business Review*, July 20, 2009, http://blogs.hbr.org/bregman/2009/07/an-18minute-plan-for-managing.html.

102. B. J. Fogg, "Three Steps to New Habits," http://www.slideshare.net/captology/3-steps-to-new-habits.

103. "Pomodoro Technique," http://www.pomodorotechnique.com.

Chapter 2

1. Available at http://www.martinlutherking.org.

2. "About the Pacific Northwest Tree Octopus," http://zapatopi.net/treeoctopus.

3. Available at http://www.thepregnancytester.com.

4. Available at http://www.archive.org/web/web.php.

5. Available at http://genochoice.com.

6. Available at http://hetracil.com.

7. "Crap Detection," http://www.diigo.com/user/hrheingold/crap_detection.

8. Richard McManus, "Detecting Bull: How to Identify Bias and Junk Journalism in Print, Broadcast, and on the Wild Web," http://www.detectingbull.com.

9. Available at http://www.easywhois.com.

10. Available at http://www.alexa.com.

11. Available at http://network-tools.com.

12. Scott Rosenberg, "In the Context of Web Context: How to Check Out Any Web Page," *Wordyard Blog*, September 14, 2010, http://www.wordyard.com/2010/09/14/in-the-context-of-web-context-how-to-check-out-any-web-page.

13. Eyder Peralta and Andy Carvin, "'Gay Girl in Damascus' Turns Out to Be an American Man," *Two-Way: NPR's News Blog*, June 13, 2011, http://www.npr.org/blogs/thetwo-way/2011/06/13/137139179/gay-girl-in-damascus-apologizes-reveals-she-was-an-american-man.

14. Jessie Daniels, "Cloaked Sites Key to Right-wing Propaganda," *Racism Review*, September 25, 2009, http://www.racismreview.com/blog/2009/09/25/cloaked-sites-key-to-right-wing-propaganda.

15. "Center for Consumer Freedom," http://en.wikipedia.org/wiki/Center_for_Consumer_Freedom.

16. "Center for Consumer Freedom," http://www.sourcewatch.org/index.php?title=Center_for_Consumer_Freedom.

17. B. J. Fogg and Hsiang Tseng, "The Elements of Computer Credibility," in Proceedings of the SIGCHI Conference on Human Factors in Computing Systems: The CHI Is the Limit, 80–87 (Pittsburgh: ACM Press, 1999).

18. B. J. Fogg, Jonathan Marshall, Othman Laraki, Alex Osipovich, Chris Varma, Nicholas Fang, Jyoti Paul, Akshay Rangnekar, John Shon, Preeti Swani, and Marissa Treinen, "What Makes Web Sites Credible? A Report on a Large Quantitative Study,"

in *Proceedings of ACM CHI 2001 Conference on Human Factors in Computing Systems*, 1:61–68 (New York: ACM Press, 2001).

19. Carolyn Y. Johnson, "Author on Leave after Harvard Inquiry," *Boston Globe*, August 10, 2010, B1.

20. Dan Barry, David Barstow, Jonathan D. Glater, Adam Liptak, and Jacques Steinberg, "Correcting the Record: *Times* Reporter Who Resigned Leaves Long Trail of Deception," *New York Times*, May 11, 2003, N1.

21. Eszter Hargittai, Lindsay Fullerton, Ericka Menchen-Trevino, and Kristin Yates Thomas, "Trust Online: Young Adults' Evaluation of Web Content, *International Journal of Communication* 4 (2010): 468–494.

22. R. David Lankes, "Trusting the Internet," *Journal of Software Technology* 14, no. 3 (August 2011), http://journal.thedacs.com/issue/58/191.

23. Lindsay Pettingill, "Trust without Knowledge: How Young Persons Carry Out Research on the Internet," GoodWork Project Report Series, no. 48 (Cambridge, MA: Harvard University, 2006), 6.

24. Soo Young Rieh and Brian Hilligoss, "College Students' Credibility Judgments in the Information-Seeking Process," in *Digital Media, Youth, and Credibility*, ed. Miriam J. Metzger and Andrew J. Flanagin, 49–72, John D. and Catherine T. MacArthur Foundation Series on Digital Media and Learning (Cambridge, MA: MIT Press, 2008).

25. Clay Shirky, "A Speculative Post on the Idea of Algorithmic Authority," 2009, http://www.shirky.com/weblog/2009/11/a-speculative-post-on-the -idea-of-algorith mic-authority.

26. Miriam J. Metzger, Andrew J. Flanagin, and Ryan B. Medders, "Social and Heuristic Approaches to Credibility Evaluation Online," *Journal of Communication* 60, no. 3 (September 2010): 413–439.

27. Mizuko Ito, Heather A. Horst, Matteo Bittanti, danah boyd, Becky Herr-Stephenson, Patricia G. Lange, C. J. Pascoe, and Laura Robinson, "White Paper—Living and Learning with New Media: Summary of Findings from the Digital Youth Project," *Digital Youth Research*, http://digitalyouth.ischool.berkeley.edu/report.

28. Mizuko Ito, interview with the author, December 17, 2009.

29. Metzger, Flanagin, and Medders, "Social and Heuristic Approaches, " 29.

30. Heuer, a social media marketing expert—a job description that would have made me smile when I started exploring BBS systems in the early 1980s—founded the Social Media Club, http://socialmediaclub.org.

31. "Basic Search Help," http://www.google.com/support/websearch/bin/static .py?hl=en&page=guide.cs&guide=1221265&answer=134479&rd=1.

32. "Even More about Dan," http://sites.google.com/site/dmrussell2/dmr-bio-info.

33. Interview with Russell on March 7, 2011.

34. Betsy Aoki, "Critical Thinking Roundup: Howard Rheingold, Bing at ISTE," *Huffington Post*, January 27, 2010, http://www.huffingtonpost.com/betsy-aoki/critical-thinking-roundup_b_642945.html.

35. "Welcome to the Critical Thinking Compendium," http://critical-thinking.iste.wikispaces.net.

36. Bing and the Microsoft Education Team, "From Search to Research: Developing Critical Thinking through Web Research Skills," http://download.microsoft.com/download/A/6/4/A645E848-4937-4564-9CF6-16A57EF8BF48/CriticalThinking.pdf.

37. "Critical Thinking in the Classroom," http://www.microsoft.com/education/en-us/teachers/guides/Pages/critical_thinking.aspx.

38. "Productivity Stats," http://chronicle.com/stats/productivity.

39. "Pop," http://www.harzing.com/pop.htm.

40. "Scholar Index," http://interaction.lille.inria.fr/~roussel/projects/scholarindex/index.cgi.

41. Available at http://factchecked.org.

42. Available at http://www.factchecked.org.

43. "Politics and Government," http://urbanlegends.about.com/od/government/Politics_and_Government.htm.

44. "Conspiracy Theories," http://www.america.gov/conspiracy_theories.html.

45. Available at http://www.lasikathome.com.

46. "Lasik at Home Hoax," http://www.usaeyes.org/lasik/library/lasik-home-hoax.htm.

47. Available at http://www.sciencedirect.com.

48. Available at http://www.hon.ch.

49. Available at https://www.hon.ch/HONcode/Plugin/Plugins.html.

50. Available at http://scienceroll.com/2009/05/20/health-information-online-how-to-check-the-quality.

51. "A User's Guide to Finding and Evaluating Health Information on the Web," http://www.mlahq.org/resources/userguide.html.

52. Susannah Fox, "Health Topics," Pew Internet and American Life Project, February 1, 2011, http://pewinternet.org/Reports/2011.HealthTopics/Part-3/Specific-disease-or-medical-problem.aspx.

53. Zachary F. Meisel, "Googling Symptoms: How it Can Help Patients and Doctors," *Time*, January 19, 2011, http://www.time.com/time/health/article/0,8599, 2043125,00.html.

54. Ibid.

55. "Guides for Patients and Consumers," http://www.effectivehealthcare.ahrq.gov/ index.cfm/guides-for-patients-and-consumers.

56. Dean Stephens, "Turning to Social Media and Search Engines for Smart Health Answers," Search Engine Watch, http://searchenginewatch.com/article/2065755/ Turning-to-Social-Media-Search-Engines-for-Smart-Health-Answers.

57. Available at http://www.medhelp.org.

58. Available at http://www.justanswer.com.

59. Available at http://www.sharecare.com.

60. Available at http://www.healthline.com.

61. Susannah Fox, "Cancer 2.0," Pew Internet and American Life Project, December 13, 2010, http://www.pewinternet.org/Reports/2010/30--Cancer-20/Report/Section -5.aspx.

62. Available at http://www.patientslikeme.com.

63. Available at http://curetogether.com.

64. Rachel Zimmerman, "Patient-Driven Social Network Refutes Study, Publishes Its Own Results," *CommonHealth*, April 26, 2011, http://commonhealth.wbur .org/2011/04/social-network-conducts-medical-research.

65. Available at http://organizedwisdom.com/home.

66. Available at http://www.talkabouthealth.com.

67. James Madison, "James Madison to W. T. Barry," *Founder's Constitution* 1, no. 18, document 35 (August 4, 1822), http//press-pubs.uchicago.edu/founders/documents/ v1ch18s35.html.

68. Jay Rosen, "The Legend of Trent Lott and the Weblogs," March 15, 2004, http:// archive.pressthink.org/2004/03/15/lott_case.html.

69. Yuki Noguchi, "Cameraphones Lend Immediacy to Images of Disaster," *Washington Post*, July 8, 2005, A16.

70. Helena Deards, "Twitter First Off the Mark with Hudson Plane Crash Coverage," *Editorsweblog*, January 19, 2009, http://www.editorsweblog.org/multimedia/2009/01/ twitter_first_off_the_mark_with_hudson_p.php.

71. "Death of Neda Agha-Soltan," Wikipedia, http://en.wikipedia.org/wiki/Death_ of_Neda_Agha-Soltan; Matthew Ingram, "It's Not Twitter or Facebook, It's the Power of the Network," *GigaOm*, January 29, 2011, http://gigaom.com/2011/01/29/twitter-facebook-egypt-tunisia.

72. Ashley Parker, "Twitter's Secret Handshake," *New York Times*, June 10, 2011, ST1.

73. Chris Messina, "Twitter Hashtags for Emergency Coordination and Disaster Relief," *FactoryCity*, October 22, 2007, http://factoryjoe.com/blog/2007/10/22/twitter-hashtags-for-emergency-coordination-and-disaster-relief.

74. "Scandalous Scoop Breaks Online," *BBC News*, January 25, 1998, http://news.bbc.co.uk/2/hi/special_report/1998/clinton_scandal/50031.stm.

75. Ben Parr, "Mindblowing #IranElection Stats: 221, 744 Tweets per Hour at Peak," *Mashable*, June 17, 2009, http://mashable.com/2009/06/17/iranelection-crisis-numbers.

76. Marc Ambinder, "Follow the Developments in Iran Like a CIA Analyst," *Atlantic*, June 15, 2009, http://www.theatlantic.com/politics/archive/2009/06/follow-the-developments-in-iran-like-a-cia-analyst/19381.

77. Dan Gillmor, *We the Media: Grassroots Journalism by the People, for the People* (Sebastopol, CA: O'Reilly Media, 2006).

78. Dan Gillmor, *Mediactive* (published by the author, 2010), http://mediactive.com.

79. Ibid., 15.

80. Cass Sunstein, "The Daily We," *Boston Review* (Summer 2001): 4–9.

81. Eli Pariser, *The Filter Bubble* (New York: Penguin, 2011).

82. Available at http://globalvoicesonline.org.

83. Gillmor, *Mediactive*, 16–24.

84. Available at http://fairspin.org.

85. Available at http://newstrust.net.

86. "Craig Shergold," http://www.snopes.com/inboxer/medical/shergold.asp.

87. M. G. Siegler, "Eric Schmidt: Every 2 Days We Create as Much Information as We Did up to 2003," *TechCrunch*, August 4, 2010, http://techcrunch.com/2010/08/04/schmidt-data.

88. Giselle Tsirulnik, "More Than 7 trillion SMS Messages Will Be Sent in 2011: ABI Research," *Mobile Marketer*, December 30, 2010, http://www.mobilemarketer.com/cms/news/research/8631.html.

89. Mike Stenger, "Facebook Serves 1 Trillion Display Ads a Year Shows Stats," *Indy-Posted*, November 8, 2010, http://www.indyposted.com/125282/facebook-serves-1-trillion-display-ads-a-year-shows-stats.

90. Roger E. Bohn and James E. Short, "How Much Information? 2009 Report on American Consumers," Global Information Industry Center, University of California at San Diego, December 2009, http://hmi.ucsd.edu/pdf/HMI_2009_Consumer-Report_Dec9_2009.pdf.

91. Denis Diderot, *Encyclopédie ou dictionnaire des sciences, des arts et des métiers, par une société, de gens de lettres*, trans. in Keith M. Baker, ed., *The Old Regime and the French Revolution* (1987), 85–86, cited in Jeremy Norman, "Diderot on Information Overload (1755)," *From Cave Paintings to the Internet*, http://www.historyofinformation.com/index.php?id=2877.

92. Ann Blair, "Strategies for Coping with Information Overload, ca. 1550–1700," *Journal of the History of Ideas* 64, no. 1 (January 2003), 11.

93. Brian W. Ogilvie, "The Many Books of Nature: Renaissance Naturalists and Information Overload," *Journal of the History of Ideas* 64, no. 1 (January 2003): 30.

94. Robin Good (Luigi Canali de Rossi), "Real-time News Curation—The Complete Guide, Part 6: The Tools Universe," *MasterNewMedia*, http://www.masternewmedia.org/real-time-news-curation-the-complete-guide-part-6-the-tools-universe.

95. Available at http://paper.li.

96. Available at http://www.aggregage.com.

97. Available at http://www.google.com/alerts.

98. Available at http://www.netvibes.com.

99. Howard Rheingold, "Meet Meredith Stewart: Teacher . . . Innovator . . . Collaborator," *DMLcentral*, December 3, 2009, http://dmlcentral.net/blog/howard-rheingold/meet-meredith-stewart-teacherinnovatorcollaborator.

100. Available at http://wikitrust.soe.ucsc.edu.

101. Available at http://en.wiki-watch.de.

102. Jesse Johan, "Aggregated Trustworthiness: How Social Dynamics Redefine Online Credibility and Help Develop the Ethos Engine." Unpublished master's thesis, IT University of Copenhagen, August 2010, 43.

103. Clay Shirky, "A Speculative Post on Algorithmic Authority," *Clay Shirky Weblog*, November 15, 2009, http://www.cjr.org/overload/interview_with_clay_shirky_par.php.

104. Russ Juskalian, "Interview with Clay Shirky, Part I," *Columbia Journalism Review*, December 19, 2008, http://www.cjr.org/overload/interview_with_Clay_Shirky_par .php.

105. Available at http://digg.com/news.

106. Ole Olson, "Digg Patriots Censorship, Part 2: The Evidence," *News Junkie*, October 15, 2010, http://newsjunkiepost.com/2010/10/15/digg-patriots-censorship -part-2-the-evidence.

107. Shirky, "A Speculative Post."

108. Ibid.

109. Lee Rainie, interview with the author, September 2010.

110. Available at http://www.sulia.com.

111. Michael G. Noll, Ching-man Au Yeung, Nicholas Gibbins, Christoph Meinel, and Nigel Shadbolt, "Telling Experts from Spammers: Expertise Ranking in Folksonomies," *Proceedings of the 32nd International ACM SIGIR Conference on Research and Development in Information Retrieval*, 612–619 (New York: ACM Press, 2009), http:// portal.acm.org/citation.cfm?id=1571941.1572046.

112. Available at http://truthy.indiana.edu/about.

113. "The Word: Wikiality," *Colbert Report*, video clip, July 31, 2006, http://www. colbertnation.com/the-colbert-report-videos/72347/july-31-2006/the-word ---wikiality.

114. Available at http://truthy.indiana.edu/about.

115. Available at http://datasift.net.

116. Available at http://zerozero88.com.

117. Available at http://ushahidi.com/products/swiftriver-platform.

Chapter 3

1. Peter Wolchak, "Meet Heather Lawver, Internet Hero," *Channelworld India*, April 5, 2001, http://www.itworldcanada.com/news/meet-heather-lawver-internet-hero/ 129365.

2. Bev Harris, "Inside a U.S. Election Vote Counting Program," *Scoop Independent News*, July 8, 2003, http://www.scoop.co.nz/stories/HL0307/S00065.htm.

3. David D. Kirkpatrick and David E. Sanger, "A Tunisian-Egyptian Link That Shook Arab History," *New York Times*, February 13, 2011, A1.

4. Lee Rainie, "Networked Creators: How Users of Social Media Have Changed the Ecology of Information" (paper presented at VALA Libraries 2010 Conference, Melbourne, February 11, 2010), http://www.pewinternet.org/Presentations/2010/Feb/~/media/Files/Presentations/2010/Feb/Lee%20Rainie%20VALA%20paper%201%20 26%2010.pdf.

5. Tim O'Reilly, The Architecture of Participation, March, June 2004, http://oreilly. com/pub/a/oreilly/tim/architecture_of_participation.html.

6. Henry Jenkins, Ravi Puroshotma, Katherine Clinton, Margaret Weigel, and Alice J. Robison, "Confronting the Challenges of Participatory Culture: Media Education for the 21st Century" (2005), http://www.newmedialiteracies.org/files/working/ NMLWhitePaper.pdf.

7. Tim O'Reilly, The Architecture of Participation, O'Reilly About, June 2004, http:// oreilly.com/pub/a/oreilly/tim/articles/architecture_of_participation.html.

8. RU Sirius, "Sheep That Shit Grass, or the End of Scarcity: Interview with Cory Doctorow," h+magazine, March 18, 2009, http://hplusmagazine.com/2009/03/18/ sheep-shit-grass-or-end-scarcity.

9. Jenkins et al., "Confronting the Challenges of Participatory Culture: Media Education for the 21st Century."

10. Ibid., 7.

11. Alvin Toffler, The Third Wave (New York: Bantam, 1980).

12. "A Whopping 35 Hours of Video Uploaded to YouTube Every Minute, 2 Billion Videos Watched Daily," HotAirPundit, March 11, 2011, http://www.hapblog .com/2011/03/whopping-35-hours-of-video-uploaded-to.html.

13. "5,000,000,000," Flickr Blog, September 19, 2010, http://blog.flickr.net/ en/2010/09/19/5000000000.

14. Henry Jenkins, interview with the author, August 26, 2010.

15. Henry Jenkins, "Critical Information Studies for a Participatory Culture," Hen ryjenkins.org, April 8, 2009, http://henryjenkins.org/2009/04/what_went_wrong_ with_web_20_cr.html.

16. Mizuko Ito, Heather A. Horst, Matteo Bittanti, danah boyd, Becky Herr-Stephenson, Patricia G. Lange, C. J. Pascoe, and Laura Robinson, Living and Learning with New Media: Summary of Findings from the Digital Youth Project (Cambridge, MA: MIT Press, 2009).

17. Ito Mizuko, "Amateur Cultural Production and Peer-to-Peer Learning" (paper presented at the annual meeting of the American Educational Research Association, New York, July 2, 2008), http://digitalyouth.ischool.berkeley.edu/node/114.

18. Ibid.

19. Mizuko Ito, "New Media and Its Superpowers: Learning, Post Pokemon" (paper presented at the National Association of Independent Schools, Denver, February 27, 2010), http://www.itofisher.com/mito/publications/new_media_and_i_1.html.

20. Ibid.

21. Ito et al., "Living and Learning with New Media."

22. Fred Turner, "Where the Counterculture Met the New Economy: The WELL and the Origins of Virtual Community," *Technology and Culture* 46, no. 3 (July 2005): 485–512, http://www.stanford.edu/~fturner/Turner%20Romantic%20Automatism% 20Journal%20of%20Visual%20Culture.pdf.

23. Howard Rheingold, *Tomorrow* (1998), http://www.rheingold.com/rants.

24. Scott Rosenberg, *Say Everything: How Blogging Began, What It's Becoming, and Why It Matters* (New York: Crown, 2009), http://www.sayeverything.com.

25. Scott Rosenberg, "Blogging, Empowerment, and the 'Adjacent Possible,'" October 8, 2010, http://www.wordyard.com/2010/10/08/blogging-empowerment-and-the-adjacent-possible.

26. Phil Agre, "Finding Your Voice: Writing for a Webzine," July 30, 1999, http://polaris.gseis.ucla.edu/pagre/zine.html.

27. Erving Goffman, *Presentation of Self in Everyday Life* (New York: Anchor, 1959).

28. Will Richardson, *Blogs, Wikis, Podcasts, and Other Powerful Web Tools for Classrooms* (Thousand Oaks, CA: Corwin, 2010).

29. Will Richardson, "Connective Writing," *Web-logged*, November 6, 2005, http://weblogg-ed.com/2005/11/06.

30. Ken Smith, "CCCC Waves and Ripples," March 30, 2004, http://www.mchron.net/site/edublog_comments.php?id=P2636_0_13_0.

31. Don Hazen, "Sinclair's Shame," *AlterNet*, October 18, 2004, http://www.alternet.org/media/20208.

32. Dave Eberhart, "How the Blogs Torpedoed Dan Rather," Newsmax.com, January 31, 2005, http://www.alternet.org/media/20208.

33. Ross Mayfield, "Power Law of Participation," *Ross Mayfield's Weblog*, April 27, 2006, http://ross.typepad.com/blog/2006/04/power_law_of_pa.html.

34. Cited in Steve Rosenbaum, "Why Curation Is Just as Important as Creation," *Mashable*, March 17, 2011, http://mashable.com/2011/03/17/curation-importance.

35. Howard Rheingold, "Virtual Communities," *Whole Earth Review* (Winter 1987), http://journals.tdl.org/jvwr/article/view/293/247.

36. Because of my long-standing interest in curation, and because I'm known as a curator myself (you can find my Diigo.com and Delicious.com bookmarks under the user name hrheingold, and I'm Johnny Shoepainter on Flickr.com, Howard Rheingold on Scoop.it, and howardrheingold on YouTube), I knew just who to talk to about curation skills.

37. Robert Scoble, interview with the author, August 11, 2010.

38. Ibid.

39. Ibid.

40. Robert Scoble, "Seven Needs of Real-time Curators," Scobleizer.com, March 27, 2010, http://scobleizer.com/2010/03/27/the-seven-needs-of-real-time-curators.

41. Ibid.

42. Available at http://masternewmedia.org.

43. Robin Good, "Real-time News Curation, Newsmastering, and Newsradars: The Complete Guide, Parts 1–6," MasterNewMedia.com, September 7, 2010, http://www.masternewmedia.org/real-time-news-curation-newsmastering-and-newsradars-the-complete-guide-part-1.

44. Ibid.

45. Ibid.

46. Ibid.

47. Ibid.

48. David Weinberger, *Everything Is Miscellaneous* (Cambridge, MA: Perseus, 2007).

49. Andrew Dermont, "The Data-Driven Economy," *Big Think*, November 17, 2010, http://bigthink.com/ideas/24807.

50. Marshall Kirkpatrick, "FreeRice: Legit or Not, It's Fun," *ReadWriteWeb*, November 7, 2007, http://www.readwriteweb.com/archives/freerice_legit_or_not_its_fun .php.

51. Clay Shirky, "Making Digital Durable: What Time Does to Categories," Long Now Foundation, Seminars about Long-term Thinking, November 14, 2005, http://longnow.org/seminars/02005/nov/14/making-digital-durable-what-time-does-to-categories.

52. Michael Wesch, "Participatory Media Literacy: Why It Matters," *Digital Ethnography @ Kansas State University*, January 3, 2009, http://mediatedcultures.net/ksudigg/?p=192.

53. Mirko Tobias Schäfer, *Bastard Culture! User Participation and the Extension of Cultural Industries* (Amsterdam: Amsterdam University Press, 2011), 22.

54. Ibid.

55. Trebor Scholz, "Introduction," Internet as Playground and Factory: A Conference on Digital Labor, http://digitallabor.org.

56. Angela Saini, "Solving the Web's Image Problem," *BBC News*, May 14, 2008, http://news.bbc.co.uk/2/hi/technology/7395751.stm.

57. Julianne Pepitone, "AOL Cuts 900 Jobs after *HuffPo* Buy," *CNNMoney.com*, March 10, 2011, http://money.cnn.com/2011/03/10/technology/aol_layoffs_arm strong/index.htm.

58. Goffman, *Presentation of Self in Everyday Life*, 209, 208, 4.

59. "Factsheet," https://www.facebook.com/press/info.php?statistics.

60. Ewen Callaway, "A Facebook Profile Can Reveal the Real You," *New Scientist*, May 26, 2009.

61. Amy L. Gonzales, and Jeffrey T. Hancock, "Mirror, Mirror on My Facebook Wall: Effects of Exposure to Facebook on Self-esteem, *Cyberpsychology, Behavior, and Social Networking*, June 24, 2010, http://www.ncbi.nlm.nih.gov/pubmed/20575709.

62. danah boyd, "Living and Learning with Social Media" (paper presented at the Penn State Symposium for Teaching and Learning with Technology, State College, April 18, 2009).

63. Fred Wilson, "Self-expression Matters," *AVC Musings of a VC in NYC*, November 13, 2010, http://www.avc.com/a_vc/2010/11/self-expression-matters.html.

64. Erick Schonfeld, "Why All the Interest in Tumblr? Try 1,540 Percent Pageview Growth," *TechCrunch*, November 12, 2010, http://techcrunch.com/2010/11/12/tumblr-1540-percent-pageview-growth.

65. Greg Marra, "Decide What the World Sees When It Searches for You," *Google Social Web Blog*, March 2, 2011, http://googlesocialweb.blogspot.com/2011/03/decide-what-world-sees-when-it-searches.html.

66. Ashley Parker, "Twitter's Secret Handshake," *New York Times*, June 10, 2011.

67. "Charity: Water," Twestival, last modified February 12, 2009, http://www.char itywater.org/twestival.

68. Larisa Epatko, "Haiti Quake Propels Use of Twitter as Disaster-Relief Tool," *PBS Newshour*, February 16, 2010, http://www.pbs.org/newshour/rundown/2010/02/haiti-quake-propels-twitter-community-mapping-efforts.html.

Chapter 4

1. Tim Berners-Lee, "The World Wide Web: A Very Short Personal History," May 7, 1998, http://www.w3.org/People/Berners-Lee/ShortHistory.html.

2. Available at http://www.worldwidewebsize.com.

3. Howard Rheingold, "Howard Rheingold on Collaboration," TED talks, February 2005, http://www.ted.com/talks/howard_rheingold_on_collaboration.html.

4. "Best of Cooperation Lectures—Key Insights from Each Lecture ," uploaded by Howard Rheingold, 2005, http://ia600506.us.archive.org/6/items/HowardRhein goldIFTF/bestofclips.mov.

5. Available at http://cooperationcommons.com.

6. David Sloan Wilson, *Evolution for Everyone: How Darwin's Theory Can Change the Way We Think about Our Lives* (New York: Delacorte, 2007), 168.

7. Robin Dunbar, "Coevolution of Neocortex Size, Group Size, and Language in Humans," *Behavioral and Brain Sciences* 16, no. 4 (1993): 691–692.

8. Ibid., 682.

9. Robert Boyd, Joseph Henrich, and Peter Richerson, "Cultural Evolution of Human Cooperation: Summaries and Findings," *Genetic and Cultural Evolution of Cooperation*, ed. Peter Hammerstein (Cambridge, MA: MIT Press in cooperation with Dahlem University Press, 2003).

10. Ibid., 383.

11. Ernst Fehr and Simon Gächter, "Altruistic Punishment in Humans," *Nature* 415 (January 2002): 137.

12. Garrett Hardin, "The Tragedy of the Commons," *Science* 162, no. 3859 (1968): 123–124.

13. Elinor Ostrom, *Governing the Commons: The Evolution of Institutions for Collective Action* (Cambridge: Cambridge University Press, 1990), 90.

14. Arthur T. Himmelman, "Collaboration for a Change: Definitions, Decision-Making Models, Roles, and Collaboration Process Guide," Himmelman Consulting, http://depts.washington.edu/ccph/pdf_files/4achange.pdf.

15. Ibid.

16. Jane McGonigal, *Reality Is Broken: Why Games Make Us Better and How They Can Change the World* (New York: Penguin, 2011), 98–99.

17. Ibid., 293.

18 Jane McGonigal, "Gaming Can Make a Better World," TED talks, February 2010, http://www.ted.com/talks/jane_mcgonigal_gaming_can_make_a_better_world .html.

19. McGonigal, *Reality Is Broken*, 277–278.

20. Mike Fahey, "Foldit Makes Protein-Folding a Game," *Kotaku*, May 8, 2008, http://kotaku.com/388753/foldit-makes-protein-folding-a-game.

21. Clive Thompson, "Clive Thompson on How Games Make Work Seem Like Play," *Wired*, March 2011, http://www.wired.com/magazine/2011/02/st_thompson_living_games.

22. McGonigal, *Reality Is Broken*, 52.

23. Douglas Thomas and John Seely Brown, "You Play World of Warcraft? You're Hired! Why Multiplayer Games May Be the Best Kind of Job Training," *Wired*, April, 2006, http://www.wired.com/wired/archive/14.04/learn.html.

24. Joi Ito, interview with the author, March 20, 2010.

25. H. G. Wells, "The World Brain: The Idea of a Permanent World Encyclopedia, in *The New Encyclopédie Française* (August1937), https://sherlock.ischool.berkeley.edu/wells/world_brain.html.

26. Howard Bloom, *Global Brain: The Evolution of Mass Mind from the Big Bang to the 21st Century* (New York: Wiley, 2001).

27. Pierre Lévy, *Collective Intelligence: Mankind's Emerging World in Cyberspace*, trans. Robert Bonnono (New York: Basic Books, 1999).

28. Pierre Lévy, interview with the author, September 3, 2010.

29. Available at http://climatecolab.org/web/guest.

30. Anita Williams Wooley, Christopher F. Chabris, Alexander Pentland, Nada Hashmi, and Thomas W. Malone, "Evidence for a Collective Intelligence Factor in the Performance of Human Groups," *Science*, September 30, 2010, http://www.sciencemag.org/content/330/6004/686.abstract.

31. Barry Wellman, Anabel Quan-Haase, Jeffrey Boase, Wenhong Chen, Keith Hampton, Isabel Isla de Diaz, and Kakuko Miyata, "The Social Affordances of the Internet for Networked Individualism," *Journal of Computer-Mediated Communication* 8, no. 3 (April 2003), http://jcmc.indiana.edu/vol8/issue3/wellman.html.

32. Andrew Slack, "Dumbledore Has Landed: 5th Flight That We Funded for Haiti Reaches Destination," HPAlliance.org, June 22, 2010, http://thehpalliance.org/2010/06/dumbledore-has-landed.

33. Kody, "EVE Online Players Organize for Japanese Relief Effort," Curse.com, March 14, 2011, http://www.curse.com/articles/eve-en-news/935973.aspx.

34. Henry Jenkins, interview with the author, September 2010.

35. Howard Rheingold, "The Art of Hosting Good Conversations Online," 1998, http://www.rheingold.com/texts/artonlinehost.html.

36. Jeff Howe, "The Rise of Crowdsourcing," *Wired*, June 2006.

37. Available at http://www.innocentive.com.

38. "Innocentive: Crowdsourcing Ideas," *Finding Petroleum*, January 10, 2011, http://www.findingpetroleum.com/n/Innocentive_crowdsourcing_ideas/2bf5f9a7.aspx.

39. Don Tapscott, and Anthony Williams, "Innovation in the Age of Mass Collaboration," Bloomberg *Business Week*, February 1, 2007, http://www.businessweek.com/innovate/content/feb2007/id20070201_774736.htm.

40. Available at http://www.threadless.com.

41. "Mechanical Turk Is a Marketplace for Work," https://www.mturk.com/mturk/welcome.

42. Jacob Colker, "New @Extraordinaries Mission: Save Lives! Help @firstaidcorps Map AED Defibrillators around the World," *SparkedBlog*, August 25, 2009, http://blog.sparked.com/2009/08/25/new-extraordinaries-mission-save-lives-help-firstaid-corps-map-aed-defibrillators-around-the-world.

43. Available at http://setiathome.berkeley.edu.

44. "Folding@home: Distributed Computing," http://folding.stanford.edu.

45. Noah Rimon, "Folding@home Petaflop Barrier Crossed," *Playstation Blog*, September 19, 2007, http://blog.us.playstation.com/2007/09/25/foldinghome-petaflop-barrier-crossed-update.

46. Jeffrey R. Young, "Crowd Science Reaches New Heights," *Chronicle of Higher Education* 56, no. 37, June 4, 2010:A14.

47. "Help Map Mars," http://beamartian.jpl.nasa.gov/maproom.

48. "About Us," http://www.genmapp.org/about.html.

49. Barend Mons et al., "Calling on a Million Minds for Community Annotation in WikiProteins," *Genome Biology* 9, R89 (2008), http://genomebiology.com/2008/9/5/R89.

50. John Markoff, "RNA Game Lets Players Help Find a Biological Prize," *New York Times*, January 10, 2011, D4.

51. Paul M. Aoki et al., "Citizen Science," Common Sense Research Project, 2008, http://www.urban-atmospheres.net/CitizenScience.

52. Eric Paulos, "Enabling Citizen Science," O'Reilly Emerging Technology Conference, March 11, 2009, http://en.oreilly.com/et2009/public/schedule/detail/5565.

53. "CrisisWiki," http://wiki.crisiscommons.org/index.php?title=Crisis_Wiki.

54. "Main Page," http://www.hurricanewiki.org/wiki/Main_Page.

55. South-East Asia Earthquake and Tsunami Blog, last modified August 11, 2009, http://tsunamihelp.blogspot.com.

56. Megha Bahree, "Citizen Voices," *Forbes*, December 8, 2008.

57. Kate Greene, "Crowdsourcing Jobs to a Worldwide Mobile Workforce," *Technology Review*, November 24, 2010, http://www.technologyreview.com/business/26651 /?p1=A5.

58. Ibid.

59. Lindsey Hoshaw, "Afloat in the Ocean, Expanding Islands of Trash," *New York Times*, November 9, 2009, D2.

60. Daniel G. Maxwell, Raphael Sofaer, and Ilya Zhitomirskiy, "Decentralize the Web with Diaspora," Kickstarter.com, last modified June 21, 2010, http://www.kick starter.com/projects/196017994/diaspora-the-personally-controlled-do-it-all-distr.

61. "The Age of Stupid," http://en.wikipedia.org/wiki/The_Age_of_Stupid.

62. Carlye Adler, "How Kickstarter Became a Lab for Daring Prototypes and Ingenious Products," *Wired*, March 18, 2011, http://www.wired.com/magazine/2011/03/ ff_kickstarter/all/1.

63. Available at http://www.donorschoose.org.

64. Ankit Sharma, "Crowdsourcing Critical Success Factor Model: Strategies to Harness the Collective Intelligence of the Crowd," Working Paper 1–2010, May 2010, http://irevolution.files.wordpress.com/2010/05/working-paper1.pdf.

65. "Apache HTTP Server," http://en.wikipedia.org/wiki/Apache_HTTP_Server.

66. Yochai Benkler, "Coase's Penguin, or Linux and the Nature of the Firm," *Yale Law Journal* (November 2002): 369.

67. Steven Weber, *The Success of Open Source* (Cambridge, MA: Harvard University Press, 2004).

68. Ibid., 73–81.

69. Rishab Aiyer Ghosh, "Cooking Pot Markets: An Economic Model for the Trade in Free Goods and Services on the Internet," *First Monday* 3, no. 3 (March 1998), http://firstmonday.org/htbin/cgiwrap/bin/ojs/index.php/fm/article/view/580/501.

70. Weber, *Success of Open Source*, 73–81.

71. Rachel Botsman and Roo Rogers, *What's Mine Is Yours: The Rise of Collaborative Consumption* (New York: Harper Business, 2010), http://collaborativeconsumption. com.

72. "Antoine de Saint-Exupery Quotes," http://thinkexist.com/quotation/if_you_ want_to_build_a_ship-don-t_drum_up_people/170927.html.

73. Lee Rainie, "Wikipedia: When in Doubt, Multitudes Seek It Out," Pew Research Center Publications, April 24, 2007, http://pewresearch.org/pubs/460/wikipedia.

74. Dan Carlin, "Corporate Wikis go Viral," *Bloomberg BusinessWeek*, March 12, 2007, http://www.businessweek.com/technology/content/mar2007/tc20070312 _476504.htm.

75. "Policing Act Launched," New Zealand Police, September 25, 2007, http://www .police.govt.nz/news/release/3370.html.

76. Available at http://www.futuremelbourne.com.au/wiki/view/FMPlan.

77. Beth Simone Noveck, "Wiki Government," *Democracy* 7 (Winter 2008): 31–43.

78. Ibid., 38.

79. John Seigenthaler, "A False Wikipedia 'Biography,'" *USA Today*, November 29, 2005, http://www.usatoday.com/news/opinion/editorials/2005-11-29-wikipedia-edit_x.htm.

80. John Timmer, "Wikipedia Hoax Points to Limits of Journalists' Research," *ars technical*, May 7, 2009, http://arstechnica.com/media/news/2009/05/wikipedia-hoax-reveals-limits-of-journalists-research.ars.

81. "The Word: Wikiality," *Colbert Report*, video clip, July 31, 2006, http://www.colbertnation.com/the-colbert-report-videos/72347/july-31-2006/the-word---wikiality.

82. "Wikipedia," http://en.wikipedia.org/wiki/Wikipedia.

83. Jim Giles, "Internet Encyclopaedias Go Head to Head," *Nature* 438 (December 15, 2005): 900–901.

84. danah boyd, "Information Access in a Network World" (paper presented at Pearson Publishing, Palo Alto, CA, November 2, 2007), http://www.danah.org/papers/talks/Pearson2007.html.

Chapter 5

1. *Six Degrees of Separation* (play), http://en.wikipedia.org/wiki/Six_Degrees_of _Separation_(play).

2. Stanley Milgram, "The Small World Problem," *Psychology Today* 2 (1967): 60–67.

3. Duncan Watts and Steven Strogatz, "Collective Dynamics of 'Small World' Networks," *Nature* 393 (June 1998): 440–442.

4. "The Oracle of Bacon," http://oracleofbacon.org.

5. Frederic Lardinois, "On Twitter, It's Just Five Degrees of Separation," April 29, 2010, http://www.readwriteweb.com/archives/six_degrees_of_seperation_on_twit ter.php.

6. Jure Leskovec and Eric Horvitz, "Planetary-Scale Views on an Instant Messaging Network," *arXiv.org*, March 6, 2008, http://arxiv.org/abs/0803.0939v1.

7. Clay Shirky, "Power Laws, Weblogs, and Inequality," *Clay Shirky's Writings about the Internet*, February 8, 2003, http://shirky.com/writings/powerlaw_weblog.html.

8. Chris Anderson, "The Long Tail," *Wired*, October 2004, http://www.wired.com/wired/archive/12.10/tail.html.

9. James H. Fowler and Nicholas Christakis, "Dynamic Spread of Happiness in a Large Social Network: Longitudinal Analysis over 20 Years in the Framingham Heart Study," *British Medical Journal* 337, no. a2338 (2008): 1–9.

10. Nicholas Christakis and James Fowler, *Connected: The Surprising Power of Social Networks* (Boston: Little, Brown, 2009).

11. Cited in Howard Rheingold, *Smart Mobs* (Cambridge, MA: Perseus, 2002): 59–60.

12. David P. Reed, "That Sneaky Exponential: Beyond Metcalfe's Law to the Power of Community Building," http://www.reed.com/dpr/locus/gfn/reedslaw.html.

13. Manuel Castells, *The Rise of the Network Society (The Information Age: Economy, Society, and Culture, Volume 1)* (London: Wiley-Blackwell, 2000).

14. Manuel Castells, "Why Networks Matter," in *Network Logic: Who Governs in an Interconnected World?* ed. Helen McCarthy, Paul Miller, and Paul Skidmore, 221–224 (London: Demos, 2004): 222–223.

15. Ibid., 224.

16. Mark Hollander, "Tools to Visualize Your Facebook Network," *Group 80/20 Mobile Medical Media Blog*, January 31, 2009, http://group8020.com/social-media/facebook-visualization-2034.

17. Cited in Howard Rheingold, *The Virtual Community* (Cambridge, MA: Perseus Books, 1993), 13.

18. Mark S. Granovetter, "The Strength of Weak Ties," *American Journal of Sociology* 78, no. 6 (May 1973): 1360–1380.

19. Ibid., 1361.

20. Marc Smith, interview with the author, April 10, 2010.

21. John Hagel III and John Seely Brown, "The Multiplier Effect," *Economist*, May 25, 2011, http://ideas.economist.com/blog/multiplier-effect.

22. Fred Turner, "Where the Counterculture Met the New Economy: The WELL and the Origins of Virtual Community," *Technology and Culture* 46, no. 3 (July 2005): 485–512.

23. Smith, interview.

24. Barry Wellman, interview with the author, September 23, 2010.

25. Barry Wellman, "The Network Community," in *Networks in the Global Village: Life in Contemporary Communities*, ed. Barry Wellman (Boulder, CO: Westview Press, 1998).

26. Barry Wellman, "Computer Networks as Social Networks," *Science* 293, no. 5537 (September 2001): 2031.

27. Ibid., 2033.

28. Lee Rainie, John Horrigan, Barry Wellman, and Jeffrey Boase, "The Strength of Internet Ties," Pew Internet and American Life Project, January 25, 2006, http://www.pewinternet.org/Reports/2006/The-Strength-of-Internet-Ties.aspx: 1.

29. Lee Rainie and Barry Wellman, *Networked: The New Social Operating System* (Cambridge, MA: MIT Press, 2012). [Quotes and paraphrases from prepublication draft.]

30. Barry Wellman, Anabel Quan-Haase, Jeffrey Boase, Wenhong Chen, Keith Hampton, Isabel Isla de Diaz, and Kakuko Miyata, "The Social Affordances of Networked Individualism," *Journal of Computer Mediated Communication* 8, no. 3 (April 2003).

31. Rainie and Wellman, *Networked*.

32. Marc Smith, "How to Build a Collection of Influential Followers in Twitter Using Social Network Analysis and NodeXL," ConnectedAction.net, March 5, 2011, http://www.connectedaction.net/2011/03/05/how-to-build-a-collection-of-influential-followers-in-twitter-using-social-network-analysis-and-nodexl.

33. Rainie and Wellman, *Networked*.

34. Ibid.

35. Ibid.

36. Ibid.

37. Ibid.

38. Ibid.

39. Gabriele Plickert, Barry Wellman, and Rochelle Côté, "It's Not Who You Know, It's How You Know Them: Who Exchanges What with Whom," http://homes.chass.utoronto.ca/~wellman/publications/reciprocity05/reciprocity3-5.pdf.

40. Barry Wellman, "Connected Lives: The New Social Operating System," Clinton School Speaker Series, April 14, 2009, http://www.clintonschoolspeakers.com/lecture/view/connected-lives-new-social-network-operating-syste.

41. Lyda J. Hanifan, "The Rural School Community Center," *Annals of the American Academy of Political and Social Science* 67 (1916): 130–138.

42. Pierre Bourdieu, *Outline of a Theory of Practice* (Cambridge: Cambridge University Press, 1972).

43. James Coleman, "Social Capital in the Creation of Human Capital," *American Journal of Sociology Supplement* 94 (1988): S95–S120.

44. Barry Wellman and Scot Wortley, "Different Strokes from Different Folks: Community Ties and Social Support," *American Journal of Sociology* 96, no. 3 (1990): 558–588.

45. Robert Putnam, *Bowling Alone: The Collapse and Revival of American Community* (New York: Simon and Schuster, 2000).

46. Lee Rainie, Kristen Purcell, and Aaron Smith, "The Social Side of the Internet," Pew Internet and American Life Project, January 18, 2011, http://pewinternet.org/Reports/2011/The-Social-Side-of-the-Internet.aspx.

47. Robert Leonardi, Rafaella Y. Nanetti, and Robert Putnam, *Making Democracy Work: Civic Traditions in Modern Italy* (Princeton, NJ: Princeton University Press, 1993).

48. Carlo M. Cipola, *Before the Industrial Revolution* (New York: W. W. Norton, 1976), 131.

49. Leonardi, Nanetti, and Putnam, *Making Democracy Work*, 174.

50. Ibid., 171.

51. Kevin Kelly, *New Rules for the New Economy: 10 Radical Strategies for a Connected World* (New York: Viking, 1998), 73.

52. Mark K. Smith, "Social Capital," in *The Encyclopedia of Informal Education*, 2000–2009, http://www.infed.org/biblio/social_capital.htm.

53. Nicole B. Ellison, Charles Steinfield, and Cliff Lampe, "The Benefits of Facebook 'Friends': Social Capital and College Students' Use of Online Social Network Sites," *Journal of Computer-Mediated Communication* 12(4) (August 2007): 1143

54. Ronald S. Burt, "Social Origins of Good Ideas," draft manuscript, January 2003, http://www.prosocia.thinkingarts.myzen.co.uk/application/workfiles/social_networking/arran_SOGI.pdf.

55. Ibid.

56. Samuel Arbesman, "Social Melting Pots Foster Technological Innovation," archived as "Social Melting Pot Makes Cities More Innovative," *New Scientist* 199, no. 2677 (October 14, 2008): 23.

57. Fred Turner, *From Counterculture to Cyberculture: Stewart Brand, the Whole Earth Network, and the Rise of Digital Utopianism* (Chicago: University of Chicago Press, 2006).

58. Fred Turner, interview with the author, July 2010.

59. Ibid.

60. Keith N. Hampton, Chul-joo Lee, and Eun Ja Her, "How New Media Affords Network Diversity: Direct and Mediated Access to Social Capital through Participation in Local Settings," *New Media and Society*, June 6, 2011, http://www.mysocial network.net/downloads/NetDiversity19-diss.pdf.

61. Turner, interview.

62. Howard Rheingold, "Shelley Terrell: Global Netweaver, Curator, PLN-Builder," DMLcentral.net, blog, October 15, 2010, http://dmlcentral.net/blog/howard-rhein gold/shelly-terrell-global-netweaver-curator-pln-builder.

63. Available at http://scholar.google.com.

64. "Publish or Perish," http://www.harzing.com/pop.htm.

65. Cory Doctorow, "How Your Creepy Ex-Co-Workers Will Kill Facebook," *InformationWeek*, November 26, 2007, http://www.informationweek.com/story/showArti cle.jhtml?articleID=204203573.

66. Nick Bilton, "Facebook Changes Privacy Settings to Enable Facial Recognition," *New York Times Bits Blog*, June 7, 2011, http://bits.blogs.nytimes.com/2011/06/07/ facebook-changes-privacy-settings-to-enable-facial-recognition.

67. Ed Oswald, "Facebook on Face Recognition: Sorry, We Screwed Up!" *PCWorld*, June 8, 2011, http://www.pcworld.com/article/229831/facebook_on_face_recogni tion_sorry_we_screwed_up.html.

68. Christopher M. Hoadley, Heng Xu, Joey J. Lee, and Mary Beth Rosson, "Privacy as Information Access and Illusory Control: The Case of the Facebook News Feed Privacy Outcry," *Electronic Commerce Research and Applications* (2009), http://tc.aca demia.edu/JoeyLee/Papers/453411/Privacy_As_Information_Access_and_Illusory_ Control_The_Case_of_the_Facebook_News_Feed_Privacy_Outcry.

69. Alexander Hotz, "Facebook Privacy: 6 Years of Controversy," *Mashable*, August 25, 2010, http://mashable.com/2010/08/25/facebook-privacy-infographic.

70. Tracy Samantha Schmidt, "Inside the Backlash against Facebook," *Time*, September 6, 2006, http://www.time.com/time/nation/article/0,8599,1532225,00.html.

71. Om Malik, "Is Facebook Beacon a Privacy Nightmare?" *GigaOM*, November 6, 2007, http://gigaom.com/2007/11/06/facebook-beacon-privacy-issues.

72. danah boyd, "Taken Out of Context: American Teen Sociality in Networked Publics" (PhD diss., University of California, Berkeley, 2008), http://www.danah .org/papers/TakenOutOfContext.pdf.

73. Henry Copeland, "Are You Also Exposing Your Private Parts to Strangers on Facebook?" *blogads*, June 8, 2011, http://blog.web.blogads.com/2011/06/08/are-you-also-exposing-your-private-parts-to-strangers-on-facebook.

74. James Nixon, "Anonymous: US Army of Fakes Tracks Facebook Users," *Thinq*, March 17, 2011, http://www.thinq.co.uk/2011/3/17/anonymous-us-army-fakes-tracks-facebook-users.

75. Nick Fielding and Ian Cobain, "Revealed: US Spy Operation That Manipulates Social Media," March 17, 2011, http://www.guardian.co.uk/technology/2011/mar/17/us-spy-operation-social-networks.

76. Alan Boswell, "How Sudan Used the Internet to Crush Protest Movement," *McClatchy*, April 6, 2011, http://www.mcclatchydc.com/2011/04/06/111637/sudans-government-crushed-protests.html.

77. Goinaz Esfandiari, "In Iran, Beware of New Facebook 'Friends,'" *Radio Free Europe/Radio Liberty*, June 8, 2011, http://www.rferl.org/content/if_youre_iranian_beware_of_new_facebook_friends/24228798.html.

78. danah boyd, "Understanding Socio-Technical Phenomena in a Web 2.0 Era" (paper presented at the Microsoft Research New England Lab Opening, Cambridge, MA, September 22, 2008), http://www.danah.org/papers/talks/MSR-NE-2008.html.

79. "The Wayback Machine," http://www.archive.org/web/web.php.

80. Kirsten Grieshaber, "Facebook Party Gets out of Control after German Girl Forgets Privacy Settings," *Huffington Post*, June 5, 2011, http://www.huffingtonpost.com/2011/06/05/facebook-party-out-of-control_n_871473.html.

Chapter 6

1. Jeffrey Rosen, *The Unwanted Gaze: The Destruction of Privacy in America* (New York: Random House, 2000).

2. Viktor Mayer-Schönberger, *Delete: The Virtue of Forgetting in the Digital Age* (Princeton, NJ: Princeton University Press, 2009).

3. Whitson Gordon, "The Always Up-to-date Guide to Managing Your Facebook Privacy," *Mashable*, June 21, 2011, http://lifehacker.com/5813990/the-always-up+to+date-guide-to-managing-your-facebook-privacy.

4. Jürgen Habermas, *The Structural Transformation of the Public Sphere: An Inquiry into a Category of Bourgeois Society*, trans. Thomas Burger (Cambridge, MA: MIT Press, 1989).

5. Theodor Adorno and Max Horkheimer, *Dialectic of Enlightenment* (Stanford, CA: Stanford University Press, 2002).

6. Robert McChesney, *Rich Media, Poor Democracy: Communication Politics in Dubious Times* (Chicago: University of Illinois Press, 1999).

7. Noam Chomsky, *Manufacturing Consent: The Political Economy of the Mass Media* (New York: Pantheon, 1988).

8. Guy Debord, *The Society of the Spectacle*, trans. Donald Nicholson-Smith (New York: Zone Books, 1994).

9. Jean Baudrillard, "Simulacra and Simulations: I. The Precession of Simulacra," trans. Sheila Faria Glaser, http://www.egs.edu/faculty/jean-baudrillard/articles/simulacra-and-simulations-i-the-precession-of-simulacra.

10. Howard Rheingold, "Habermas Blows Off Question about the Internet and the Public Sphere," *SmartMobs*, November 5, 2007, http://www.smartmobs.com/2007/11/05/habermas-blows-off-question-about-the-internet-and-the-public-sphere.

11. Pieter Boeder, "Habermas' Heritage: The Future of the Public Sphere in the Network Society," *First Monday* 10, no. 9 (September 2005), http://firstmonday.org/htbin/cgiwrap/bin/ojs/index.php/fm/article/view/1280/1200.

12. Howard Rheingold, "Howard Rheingold's Public Sphere in the Internet Age Widget," *Howard Rheingold's Posterous*, February 6, 2009, http://howardrheingold.posterous.com/howard-rheingolds-public-spher.

13. Lawrence Lessig, *Remix: Making Art and Commerce Thrive in the Hybrid Economy* (New York: Penguin, 2008).

14. Howard Rheingold, *Smart Mobs* (Cambridge, MA: Perseus Books, 2002).

15. Lawrence Lessig, *Free Culture: How Big Media Uses Technology and the Law to Lock Down Culture and Control Creativity* (New York: Penguin, 2004).

16. Lewis Hyde, *Common as Air: Revolution, Art, and Ownership* (New York: Farrar, Straus and Giroux, 2010).

17. Aram Sinnreich, *Mashed Up: Music, Technology, and the Rise of Configurable Culture* (Amherst: University of Massachusetts Press, 2010).

18. Jonathan Zittrain, *The Future of the Internet—and How to Stop It* (New Haven, CT: Yale University Press, 2008).

19. danah boyd, "Taken Out of Context: American Teen Sociality in Networked Publics" (PhD diss., University of California at Berkeley, 2008), http://www.danah.org/papers/TakenOutOfContext.pdf.

20. Mizuko Ito, Heather A. Horst, Matteo Bittanti, danah boyd, Becky Herr-Stephenson, Patricia G. Lange, C. J. Pascoe, and Laura Robinson, "White Paper—Living and

Learning with New Media: Summary of Findings from the Digital Youth Project," *Digital Youth Research*, http://digitalyouth.ischool.berkeley.edu/report.

21. Linda Fogg Phillips and B. J. Fogg, *Facebook for Parents* (Stanford, CA: Captology Media, 2010).

22. Sonia Livingstone, *Children and the Internet* (Cambridge, UK: Polity, 2009).

23. danah boyd, interview with the author, June 15, 2010.

24. Cited in Betsy Morris, "Steve Jobs Speaks Out," *CNNMoney*, March 7, 2008, http://money.cnn.com/galleries/2008/fortune/0803/gallery.jobsqna.fortune/6.html.

25. "Escape Your Search Engine Filter Bubble: An Illustrated Guide," DuckDuckGo. com, http://dontbubble.us.

26. Rachel Botsman and Roo Rogers, *What's Mine Is Yours: The Rise of Collaborative Consumption* (New York: Harper Business, 2010).

27. boyd, "Taken Out of Context."

28. Douglas Rushkoff, *Program or Programmed: Ten Commands for a Digital Age* (New York: OR Books, 2010).

Index